Investigation in the Creep-Fatigue Coupled Effect of Rock Salt

Jinyang Fan · Zongze Li · Chunhe Yang · Tongtao Wang

Investigation in the Creep-Fatigue Coupled Effect of Rock Salt

Authors
See next page

ISBN 978-981-96-5430-7   ISBN 978-981-96-5431-4 (eBook)
https://doi.org/10.1007/978-981-96-5431-4

© The Editor(s) (if applicable) and The Author(s) 2025. This book is an open access publication.

Open Access This book is licensed under the terms of the Creative Commons Attribution-NonCommercial-NoDerivatives 4.0 International License (http://creativecommons.org/licenses/by-nc-nd/4.0/), which permits any noncommercial use, sharing, distribution and reproduction in any medium or format, as long as you give appropriate credit to the original author(s) and the source, provide a link to the Creative Commons license and indicate if you modified the licensed material. You do not have permission under this license to share adapted material derived from this book or parts of it.

The images or other third party material in this book are included in the book's Creative Commons license, unless indicated otherwise in a credit line to the material. If material is not included in the book's Creative Commons license and your intended use is not permitted by statutory regulation or exceeds the permitted use, you will need to obtain permission directly from the copyright holder.

This work is subject to copyright. All commercial rights are reserved by the author(s), whether the whole or part of the material is concerned, specifically the rights of translation, reprinting, reuse of illustrations, recitation, broadcasting, reproduction on microfilms or in any other physical way, and transmission or information storage and retrieval, electronic adaptation, computer software, or by similar or dissimilar methodology now known or hereafter developed. Regarding these commercial rights a non-exclusive license has been granted to the publisher.

The use of general descriptive names, registered names, trademarks, service marks, etc. in this publication does not imply, even in the absence of a specific statement, that such names are exempt from the relevant protective laws and regulations and therefore free for general use.

The publisher, the authors and the editors are safe to assume that the advice and information in this book are believed to be true and accurate at the date of publication. Neither the publisher nor the authors or the editors give a warranty, expressed or implied, with respect to the material contained herein or for any errors or omissions that may have been made. The publisher remains neutral with regard to jurisdictional claims in published maps and institutional affiliations.

This Springer imprint is published by the registered company Springer Nature Singapore Pte Ltd.
The registered company address is: 152 Beach Road, #21-01/04 Gateway East, Singapore 189721, Singapore

If disposing of this product, please recycle the paper.

Jinyang Fan
State Key Laboratory of Coal Mine Disaster
Dynamics and Control
School of Resources and Safety
Engineering
Chongqing University
Chongqing, China

Chunhe Yang
State Key Laboratory of Coal Mine Disaster
Dynamics and Control
School of Resources and Safety
Engineering
Chongqing University
Chongqing, China

State Key Laboratory of Geomechanics
and Geotechnical Engineering Safety
Institute of Rock and Soil Mechanics
Chinese Academy of Sciences
Wuhan, China

Zongze Li
State Key Laboratory of Coal Mine Disaster
Dynamics and Control
School of Resources and Safety
Engineering
Chongqing University
Chongqing, China

Division of Mining and Geotechnical
Engineering
Luleå University of Technology
Luleå, Sweden

Tongtao Wang
State Key Laboratory of Geomechanics
and Geotechnical Engineering Safety
Institute of Rock and Soil Mechanics
Chinese Academy of Sciences
Wuhan, China

Preface

For decades, research on the creep and fatigue behaviors of rock salt has been crucial for understanding the stability of subsurface energy storage systems, such as compressed air energy storage and nuclear waste repositories. The long-term performance of these systems depends significantly on the ability of rock salt to withstand complex and alternating stress conditions. Rock salt is known for its unique rheological properties, self-healing capabilities, and low permeability, making it an excellent candidate for energy storage applications. However, predicting its deformation and failure under creep-fatigue interactions remains a significant challenge. This challenge stems from the intricate coupling between time-dependent deformation (creep) and repeated loading-induced damage (fatigue), which necessitates a comprehensive investigation of the underlying mechanisms.

This book, *Investigation in the Creep-Fatigue Coupled Effect of Rock Salt*, systematically explores the mechanical behavior of rock salt under combined creep and fatigue loads. It introduces theoretical frameworks, experimental results, and constitutive models developed to capture the complex interactions between these two deformation mechanisms. By integrating laboratory experiments and numerical model, this work provides novel insights into the damage evolution and long-term stability of rock salt formations.

The book begins with an overview of rock salt's mechanical properties and their relevance to subsurface energy storage. It then details the experimental studies conducted under varying stress conditions to understand the material's response to creep-fatigue loading. The results are used to propose and validate new constitutive models that account for the observed deformation and failure characteristics. Finally, this book discusses the limitations of existing research and outlines future research directions.

This publication is intended as a resource for graduate students, researchers, and engineers in the fields of rock mechanics, geotechnical engineering, and energy storage. The research presented in this book is based on extensive work by the authors, which we hope will contribute to advancements in energy storage safety and efficiency. We have referenced significant literature in the field and provided an up-to-date perspective on the ongoing research. Despite our best efforts, there may still be gaps or errors, and we welcome constructive feedback from our readers.

Chongqing, China
Jinyang Fan
Zongze Li
Chunhe Yang
Tongtao Wang

Acknowledgements This book is supported by the National Key R&D Program of China (No. 2024YFB4007100), Deep Earth Probe and Mineral Resources Exploration-National Science and Technology Major Project (No. 2024ZD1004107), National Natural Science Foundation of China (No. 52274073). We are grateful to the State Key Laboratory of Coal Mine Disaster Dynamics and Control, School of Resources and Safety Engineering, Chongqing University, which provide us excellent work and researching environment.

We profit from the stimulating discussions and the helpful and open atmosphere in our research group constituted by talented master and Ph.D. students. In particular, our gratitude goes to Dr. Fan Yang, Dr. Pengyu Guo and Dr. Marion Fourmeau who have contributed a great deal to the final form of this book. We would also like to express our gratitude to organizations for permitting us to reproduce some of the figures.

Our final thank goes to the publisher Springer Nature Singapore Pte. Ltd. We acknowledge the excellent support of Wayne Hu (Publishing Editor) while working on the book manuscript.

Competing Interests The authors have no competing interests to declare that are relevant to the content of this manuscript.

Contents

1	**Introduction**		1
	1.1 Background and Significance		1
		1.1.1 The Necessity of Renewable Energy Usage	1
		1.1.2 Current Status of Global Energy Storage Technology Development	4
		1.1.3 Significance of the Study of Creep and Fatigue Mechanical Properties of Rock Salts	10
	1.2 State-of-the-Art		11
		1.2.1 Fatigue Mechanical Properties of Rocks Under Cyclic Loading	11
		1.2.2 Creep Mechanical Properties of Rocks Under Constant Loads	14
		1.2.3 Advancements in Rock Creep–Fatigue Mechanics Research	19
	1.3 Research Content		22
	References		23
2	**Dilatancy Properties of Salt Under Monotonous Compression and Brief Introduction into Dislocation Theory**		33
	2.1 Experimental Materials and Methods		33
		2.1.1 Rock Salt Material and Specimens	33
		2.1.2 Testing Equipment	37
		2.1.3 Test Procedure	38
	2.2 Dilatancy Features in Uniaxial Tests		38
		2.2.1 Loading Features	38
		2.2.2 Volume Expansion Features	39
		2.2.3 Elastic Constants	41
	2.3 Dilatancy in Triaxial Compression Test		43

	2.4	Dislocation Theory	47
		2.4.1 Conceptual Framework of Dislocation	47
		2.4.2 Dislocation Behavior of Salt Under Monotonous Compression	50
	2.5	Conclusions	52
	References		52
3	**Conventional Creep and Fatigue Mechanical Properties of Rock Salt**		**55**
	3.1	Experimental Methods	55
	3.2	Creep Mechanical Properties of Rock Salt Under Different Stress Levels	57
		3.2.1 Stress–Strain Curves of Rock Salt Under Step-Up/Step-Down Stress Levels in Creep Tests	57
		3.2.2 The Effect of Stress Level Changes on the Creep Mechanical Properties of Rock Salt	58
	3.3	Fatigue Mechanical Properties of Rock Salt Under Different Loading and Unloading Rates	60
		3.3.1 The Effect of Mixed-Rate Loading on the Stress–Strain Curve and Residual Strain of Rock Salt	60
		3.3.2 The Impact of Stress Levels on the Stress–Strain Curve and Residual Strain of Rock Salt	61
		3.3.3 Quantitative Relationship Between Mixed Rate Loading and the Mechanical Properties of Rock Salt	64
		3.3.4 Rate Effect Equation of Rock Salt Influenced by Stress Loading Rate	67
		3.3.5 Deformation Analysis of Rock Salt During Stress Loading Process	70
	3.4	Conclusions	72
	References		73
4	**Discontinuous Fatigue Mechanical Properties for Rock Salt**		**75**
	4.1	Experimental Methods	75
	4.2	Test Results and Analysis	76
		4.2.1 Stress–Strain Curves	76
		4.2.2 Residual Strains During in the Rock Salt Test	76
		4.2.3 Elastic Constants from the Rock Salt Test	78
		4.2.4 Time Interval Effect on Rock Salt	79
		4.2.5 Rupture Form	82
	4.3	Long Interval Effect on Rock Salt	84
		4.3.1 Experiment Setup	84
		4.3.2 Experimental Results in the Long Interval Test of Rock Salt	86

	4.4	Discontinuous Fatigue Life Model of Rock Salt	86
	4.5	Conclusions	88
	References	89
5	**Creep–Fatigue Mechanical Characterization of Rock Salt Under Uniaxial Stresses**		91
	5.1	Experimental Methods	91
	5.2	Test Results and Analysis	93
		5.2.1 Stress–Strain Curve in the Uniaxial Creep–Fatigue Tests	93
		5.2.2 Creep–Fatigue Strain Rate in Rock Salt	95
		5.2.3 Creep–Fatigue Residual Strain in Rock Salt	97
		5.2.4 Relation Between Fatigue Life and Creep Life in the Creep–Fatigue Test of Rock Salt	99
	5.3	Mechanisms of Creep–Fatigue Interactions in Rock Salts	103
	5.4	Conclusions	110
	References	110
6	**Creep–Fatigue Mechanical Characterization of Rock Salt Under Triaxial Stresses**		115
	6.1	Experimental Methods	115
		6.1.1 Triaxial Creep–Fatigue Procedure	115
		6.1.2 Test Procedure	117
	6.2	Test Results and Analysis	118
		6.2.1 Stress–Strain Curve in the Triaxial Creep–Fatigue Tests	118
		6.2.2 Impact of Confining Pressure on Creep Deformation in Rock Salt in the Triaxial Creep–Fatigue Tests	120
		6.2.3 Impact of Confining Pressure on Residual Strain in Rock Salts During the Triaxial Creep–Fatigue Tests	124
	6.3	Analysis of Confining Pressure Effects on Creep–Fatigue Properties of Rock Salts	125
		6.3.1 Influence of the Increase of the Confining Pressure on the Transformation of Brittle Ductility in Rock Salts	125
		6.3.2 Mechanical Interpretation of Rock Salts Affected by Confining Pressure and the Effect of Burial Depth on the Deformation of Surrounding Rock of the Salt Cavern	128
	6.4	Conclusions	133
	References	133

7 Multi-stage Amplitude Creep–Fatigue Mechanical Characterization of Rock Salt with Acoustic Emission Signal Analysis 137
   7.1   Experimental Methods 138
         7.1.1   Acoustic Emission System 138
         7.1.2   Data Collected by AE 140
         7.1.3   Mechanical Tests Series with AE 141
   7.2   Results and Analysis of Rock Salt Under Multistage Creep–Fatigue Loading 143
         7.2.1   Stress–Strain Curve of Rock Salt in the U/TSCF Tests 143
         7.2.2   Creep Fatigue Residual Strain of Rock Salt in U/TSCF Test 145
         7.2.3   AE Counts and Energy During Creep–Fatigue Testing of Rock Salt 146
         7.2.4   AE Peak Frequency During the Creep–Fatigue Tests of Rock Salt 148
         7.2.5   RA and AF Metrics of Rock Salt During U/TSCF Tests 151
         7.2.6   Damage Variables Based on AE Counts of Rock Salt During the U/TSCF Tests 152
   7.3   Analysis of Damage Evolution Characterization for Creep–Fatigue Properties of Rock Salts 155
         7.3.1   Discussion on Rock Failure Prediction Based on RA and AF of Rock Salts 155
   7.4   Conclusions 157
   References 158

8 Long-Time Creep–Fatigue Mechanical Properties of Rock Salt 161
   8.1   Experimental Methods 161
   8.2   Results and Analysis of Long-Time Creep–Fatigue Mechanical Properties of Rock Salt 163
         8.2.1   Stress–Strain Curve of Rock Salt Specimen 163
         8.2.2   Strain in Rock Salt at Different Loading Stages 164
         8.2.3   Strain Rate and Elastic Modulus Analysis of Creep–Fatigue Curves 167
   8.3   Mechanistic Analysis of Rock Salt Deformation Variations Due to Loading Rates 173
   8.4   Conclusions 175
   References 176

9 New Creep–Fatigue Constitutive Modeling of Rock Salt Based on State Variables ... 177
9.1 Constitutive Modeling of Rock Salt Considering State Variables ... 178
9.1.1 Typical Rock Creep Models ... 178
9.1.2 Definition of Plasticity Factor (State Variable) ... 180
9.2 Validation of the Creep–Fatigue Constitutive Model for Rock Salt in Creep, Fatigue and Creep–Fatigue Test ... 187
9.2.1 Validation of the Creep–Fatigue Constitutive Model in Pure Creep Test for Rock Salt ... 187
9.2.2 Validation of the Creep–Fatigue Constitutive Model in Pure Fatigue Test for Rock Salt ... 192
9.2.3 Validation of the Creep–Fatigue Constitutive Model in Creep–Fatigue Test for Rock Salt ... 197
9.3 Analysis of the Influence of Model Parameters in Creep and Fatigue Tests for Rock Salt ... 201
9.3.1 Analysis of the Influence of Model Parameters in Creep Tests for Rock Salt ... 201
9.3.2 Analysis of the Influence of Model Parameters in Fatigue Tests for Rock Salt ... 205
9.4 Conclusions ... 208
References ... 210

10 Conclusion ... 213
10.1 Main Conclusions and Innovations ... 213
10.2 Implications for Future Study ... 214

Chapter 1
Introduction

1.1 Background and Significance

1.1.1 The Necessity of Renewable Energy Usage

On November 6, 2022, the World Meteorological Organization (WMO) of the United Nations published the report 'Interim Global Climate 2022' (WMO 2022). According to the study, the past eight years have been the hottest on record, with rising greenhouse gas concentrations and accumulated heat contributing to the phenomenon [1]. In addition, 2022 extreme heat waves, droughts and devastating floods have affected millions of lives and caused billions of dollars in damages [2].

The signs and impacts of climate change are also becoming more pronounced. The rate of sea level rise has doubled since 1993. It has risen by nearly 10 mm since January 2020, accounting for 10% of the overall sea level rise [3]. To add insult to injury, the rate of warming is likewise accelerating, with the 10-year average temperature for the period 2013–2022 estimated to be 1.14 °C above the pre-industrial baseline of 1850–1900. This exceeds the Intergovernmental Panel on Climate Change's (IPCC) Sixth Assessment Report estimate of 1.09 °C [4]. This sequence of extreme weather events suggests that global warming caused by carbon dioxide from the world's industrial activities is already having a severe impact on the climate. Carbon dioxide-induced global warming is already having a serious impact on our living environment. This is despite the fact that at the 21st United Nations Climate Conference in 2015, 195 countries from around the world signed the *Paris Agreement*, which aims to reduce carbon dioxide emissions and minimize the effects of greenhouse gases [5, 6]. However, the climate and environmental disasters show that human beings have not significantly recognized the harm caused by climate change.

In fact, as early as 1980, the world's three major environmental protection organizations, The International Union for Conservation of Nature (IUCN), the United Nations Environment Programme (UNEP) and the World Wildlife Fund (WWF), jointly published the *World Conservation Strategy 1980* for the protection of the world's environment [7]. The concept of sustainable development was clearly put forward [8]. One of the core requirements of sustainable development is to reduce the use of non-renewable energy sources (fossil energy), accelerate the development of renewable (green) energy sources, and gradually increase their share in energy use [9]. The *Paris Agreement*, for its part, specifies targets for countries to increase the use of renewable energy to reduce greenhouse gas emissions, to limit the increase in global average temperature to less than 2 °C above pre-industrial levels and to work towards limiting the temperature increase to 1.5° C above pre-industrial levels, and calls on member States to phase out the use of coal by 2050–2100 and to achieve offsets for global carbon dioxide releases and sequestration [10].

Although since the beginning of the twentieth century, countries around the world have gradually recognized the need for renewable energy use, especially after 2015, the proportion of renewable energy in primary energy consumption has been increasing from less than 1% in 1995 to 5.7% in 2020, as required by the *Paris Agreement*. However, compared to the primary energy consumption ratios of 31.2% for oil, 27.2% for coal, and 24.7% for natural gas (Fig. 1.1), the use of renewable energy is still far from adequate [11].

In the Global Energy Transition 2050 report by the International Renewable Energy Agency (IRENA), it is pointed out that government plans still fall far short of meeting emission reduction requirements. Under current and proposed policies, there is only a possibility of less than 70% (66%) to achieve the global target of limiting temperature rise to below 2 °C [12, 13]. This calls for us to increase the scale of renewable energy deployment by at least six times. This is particularly important for China, as it currently ranks only seventh among the top 10 countries in terms of the share of renewable energy in primary energy consumption (Fig. 1.2), and has not yet reached the global average for sustainable energy use [14].

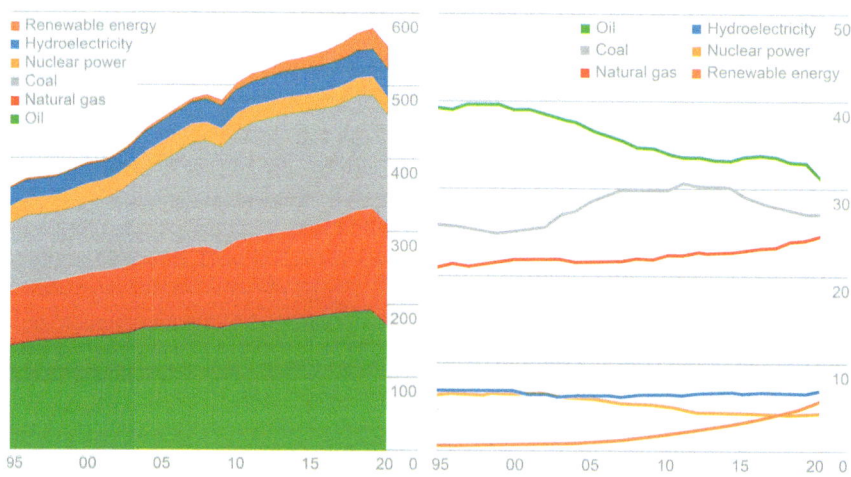

Fig. 1.1 Global energy consumption and share of primary energy consumption

1.1 Background and Significance

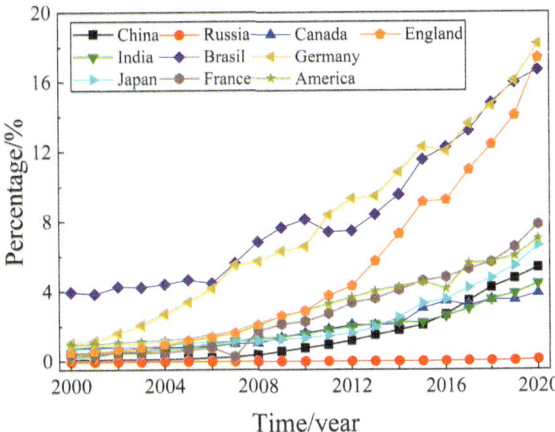

Fig. 1.2 Share of non-water renewable energy in major 10 countries all over the word

There are two main ways to accelerate the achievement of carbon peaking and carbon neutrality: (1) using carbon offset mechanisms to reduce the total amount of carbon emissions generated. Examples include decreasing the total production, afforestation or the purchase of renewable energy vouchers; (2) increase the use of renewable energy, with the ultimate goal of using only low-carbon energy sources instead of fossil fuels, so that the amount of carbon released and absorbed back into the Earth does not increase to reach equilibrium [15]. However, due to the intermittent, random and unstable nature of renewable energy (such as wind, solar, etc.), direct integration into the grid will have many adverse effects [16, 17]. As a result, renewable energy sources face difficulties in grid integration and consumption. The phenomenon of large-scale wind and light abandonment has appeared. According to statistics between 2011 and 2016, China's average annual wind abandonment rate was around 13%, which is equivalent to the energy generated by 10 million tons of standard coal (Fig. 1.3). By 2018 China's renewable energy wasted due to grid integration and consumption problems still exceeded 50 billion kWh [18, 19]. How to effectively solve the grid-connected consumption of renewable energy has become an important task to ensure the future energy security and development of the world.

In addition to constructing cross-regional power transmission channels and researching cross-regional renewable energy power scheduling technologies to improve the efficiency of renewable energy consumption [20, 21], another effective approach is to use energy storage technology for "peak shaving and valley filling", proper to renewable energy [23–25]. This approach not only addresses the large-scale utilization of renewable energy and improves its efficiency but also gradually promotes its replacement of fossil fuels as the new primary energy source. Moreover, it enhances the peak-shifting capability of the power system, making it an indispensable technological means to facilitate the global power system's comprehensive transition towards low-carbon transformation. Therefore, it is urgent to intensify the development and accelerate the construction of electricity storage projects [27, 28].

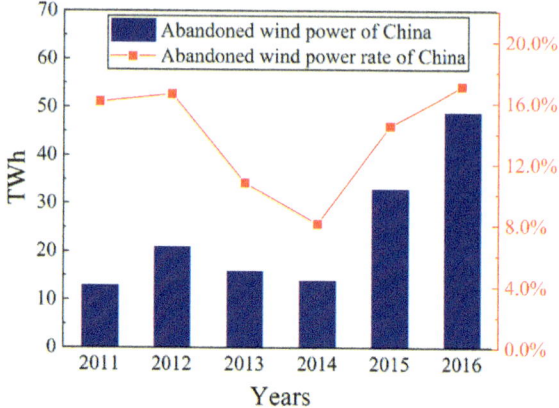

Fig. 1.3 Abandoned wind power and rate of China between 2011 and 2016

1.1.2 Current Status of Global Energy Storage Technology Development

Electricity storage technology refers to the electricity generated by excess or intermittent renewable energy sources in the trough of electricity consumption storage, in the peak of electricity consumption stable release, to meet a continuous and stable power supply technology [29, 30]. Energy storage technology can separate power generation and power consumption in time and space in two dimensions, to make up for the power system cannot be 'storage and release' shortcomings, so that the rigid power supply system becomes more flexible. This can calm the volatility of large-scale clean energy generation access to the grid, improve the security of grid operation, economy and flexibility. Energy storage is the key to guarantee the large-scale development of clean energy and the safe and economic operation of the power system [32].

Energy storage technologies can be divided into two main categories based on their storage methods: direct energy storage [33] and indirect energy storage [34]. Direct energy storage refers to the process of charging and discharging electrical energy directly through the conversion of magnetic and electric fields. It mainly includes superconducting magnetic energy storage [35] and supercapacitor energy storage [36]. These energy storage devices typically have short storage times, allowing for rapid energy release. While they have a relatively small storage capacity, they boast high energy utilization efficiency, making them suitable for short-duration and high-frequency energy storage. They are commonly used to control voltage fluctuations in power transmission networks, ensuring the safe operation of the electrical grid. Indirect energy storage differs from direct energy storage in that it involves converting surplus electrical energy before storing it. This category includes chemical energy storage [37, 38] (such as flow batteries [39], lithium/lead batteries [40]) and mechanical energy storage [41] (like flywheels [42], compressed air [43], and pumped hydro storage [44]). Among these, lithium/lead batteries and flywheel energy storage also

1.1 Background and Significance

fall under short-duration and high-frequency energy storage. Considering the characteristics of clean energy generation, such as wind and solar power, long-duration and large-scale energy storage represented by flow batteries, compressed air energy storage, and pumped hydro storage, can leverage their long cycles and high capacity to regulate the fluctuations in renewable energy generation over extended periods. This ensures stable power supply, reduces overall electricity costs, and meets the demands of the electricity grid over longer timeframes. Table 1.1 provides an overview of the main energy storage solutions, with main characteristics and pros and cons.

As of 2021, the global installed capacity of energy storage has exceeded 209 GW. Among this, the United States accounts for 34%, China accounts for 24%, and Europe accounts for 22%, collectively representing over 90% of the total [45, 46]. Pumped hydro storage, being the most mature energy storage technology, constitutes nearly 90% of the share. However, despite its prominence, pumped hydro storage faces strict limitations due to geographical potential constraints. It suffers from slow start-up, long construction periods, limited resource endowment, and relatively high costs. Although compressed air energy storage exhibits relatively lower efficiency, it offers significant advantages such as shorter construction periods, fewer storage site limitations, longer lifespan, environmental cleanliness without pollution, and unrestricted energy storage cycles. As a result, it is considered the primary direction for future energy storage technology development [48].

Existing compressed air energy storage (CAES) plants can be categorized into two types: the supplementary combustion CAES [49] and the non-supplementary combustion CAES [50], based on whether additional fossil energy is required. Both types of CAES utilize electrical energy to compress air into high-pressure gas for storage. The main difference lies in the release of high-pressure gas for power generation. The supplementary combustion CAES involves heating the compressed air during expansion to generate greater thrust and maintain system circulation. This process requires the burning of coal or natural gas to heat the air, which is known as "supplementary combustion," as shown in Fig. 1.4. As a result, traditional supplementary combustion CAES faces technological constraints, including the reliance on fossil fuels like natural gas to provide heat, leading to lower system efficiency. Typically, the overall efficiency is only around 40–55% [51].

Another type of non-supplementary combustion CAES aligns better with clean and environmentally friendly characteristics, offering promising prospects, as illustrated in Fig. 1.5. The non-supplementary combustion CAES utilizes its "internal circulation" to store the energy generated during the compression of air through heat exchange. This stored thermal energy is released during power generation, serving as a natural "propellant." The entire process involves no combustion or emissions, making it highly suitable for carbon neutrality requirements and achieving zero emissions. Moreover, it boasts higher efficiency, with electrical energy conversion rates exceeding 65% [52].

Rock salt possesses several excellent properties, including low porosity, low permeability, self-healing of damage, good rheological behavior, solubility in water for easy extraction, and chemical stability. Rock salt formations are also the most structurally dense and tightly sealed geological bodies in nature [53–55]. This makes

Table 1.1 Efficiency, cost of generation, lifetime, advantages and disadvantages of various energy storage solutions (reference)

Categorization	Type of energy storage	Efficiency/%	Power generation cost/$ Kwh	Life/year	Advantages	Disadvantages
Mechanical energy storage	Compressed air energy storage	50–70	0.45–0.5	60	Environmentally friendly and non-polluting, high efficiency	Difficulty in site selection
	Pumped hydro storage	70–75	0.2–0.26	40–50	High energy storage, long duration of energy release, mature and reliable technology	Geological conditions Demanding
	Gravitational energy storage	> 85	0.5	30	Simple principle, low technical threshold	Low energy density Excessive construction scale
	Flywheel energy storage	> 90	0.8	20	High efficiency, high instantaneous power, fast response time	Low energy density Short release time
Electrochemical energy storage	Lithium Ion battery	85–98	0.62–0.82	> 5000	High energy density, long cycle life, good power characteristics, fast response time	High prices There are security risks
	Sodium-ion battery	> 80	0.65–0.85	> 50,000	As above	As above
	Lead-acid battery	70–90	0.61–0.83	3000	Higher specific power, better safety, lower cost	Have a shorter lifespan Pollution of the environment
	Vanadium flow battery	75–85	0.7–0.95	10,000	Long battery life; power and capacity independent design; good safety	Low energy density Low reliability
Chemical energy storage	Hydrogen energy storage	30–50	> 1	1 ~ 2	Adequate supply, non-toxic and non-polluting	Expensive Low efficiency
Thermal energy storage	Molten salt energy storage	< 60	0.9	25	High energy storage density, good stability, inexpensive and easy to obtain	Low thermal conductivity, corrosive, with certain safety hazards

1.1 Background and Significance

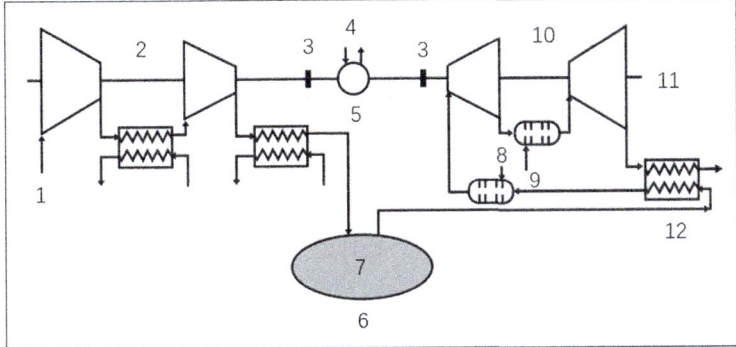

1.Air, 2.Compactors, 3.Clutches, 4. electronic, 5.electric motor/ generator, 6.Air storage 7.Compressed air, 8.Fuel, 9.Combustion chamber, 10.Turbin, 11. Exhaust gas, 12. Heat exchange

Fig. 1.4 Schematic diagram of a supplementary combustion compressed air energy storage plant

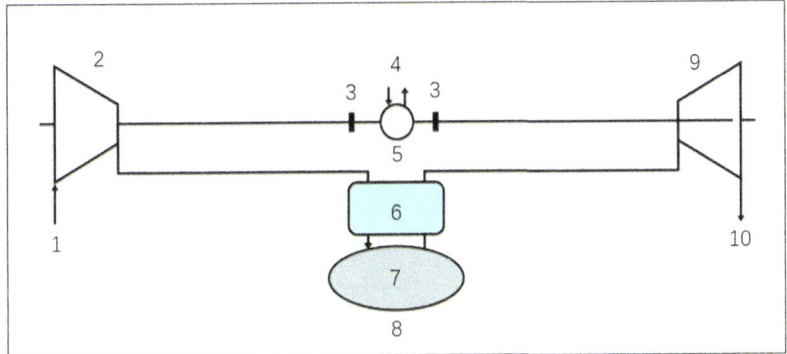

1.Air, 2.Compactors, 3.Clutches, 4. electronic, 5.electric motor/ generator, 6. Thermal storage 7.Compressed air, 8.Air storage, 9.Turbin, 10. Exhaust gas

Fig. 1.5 Schematic diagram of a non-supplementary combustion compressed air energy storage plant

the salt caverns formed after water dissolution have excellent sealing and stability, making them ideal places for storing various types of energy resources [57, 58].

In the 1950s, the United Kingdom made its first attempt to use salt caverns for oil storage. As of 2018, the United States has established over 60 salt cavern oil storage facilities, with its strategic petroleum reserves (720 million barrels) stored entirely in salt cavities. Germany, France, Russia, Canada, and other countries have also built rock salt oil storage facilities [59–61]. In terms of natural gas storage, countries utilizing salt caverns for this purpose include the United States, Canada, Germany, France, and Italy. Currently, there are more than 70 salt cavern gas storage facilities worldwide, with a total working gas capacity exceeding 25 billion cubic meters. The United States alone has 31 rock salt gas storage facilities with a total working gas

Fig. 1.6 View of the Huntorf CAES plant in Germany

capacity of 1.78 billion cubic meters [63, 64]. For compressed air energy storage, Germany built its first 290 MW salt cavern air-compressed energy storage power station in Huntorf in 1978, as shown in Fig. 1.6. It has been operating steadily for nearly 40 years [66].

In 1991, the United States established the world's second commercial salt cavern compressed air energy storage (CAES) power station in McIntosh, Alabama, with a capacity of 110 MW [67]. China's evaluation and utilization of salt cavern storage began in the 1990s. In 2004, PetroChina started construction of China's first salt cavern gas storage facility, the Jintan Salt Cavern Gas Storage, as part of the West–East Gas Pipeline project. Currently, three storage facilities have been completed, with a commercial gas supply capacity of 600 million cubic meters. During the 13th Five-Year Plan period, China also planned multiple salt cavern gas storage facilities in Jintan, Huaian, Pingdingshan, Yunying, and Zhangshu, with a total working gas capacity exceeding 8 billion cubic meters once completed. In terms of salt cavern oil storage, China has conducted extensive preliminary research, forming pre-feasibility reports, and has currently planned six salt cavern oil storage facilities [68, 69]. China's first salt cavern CAES power station, the Jintan Salt Cavern Compressed Air Energy Storage National Pilot Demonstration Project, began construction in 2017 and successfully connected to the grid for power generation on September 30, 2021 [71]. The initial phase of the power station has a generating capacity of 60 MW and a storage capacity of 300 MW/h, with a future planned capacity of 1000 MW.

France is also one of the earliest countries to establish energy reserves [72, 73]. With extensive experience in the construction, use, and management of salt caverns, it stands out with its diverse range of stored products, including not only crude oil but also various refined petroleum products. This valuable experience can provide substantial support for global strategic underground energy storage in salt caverns. France has completed over 10 operational salt cavern storage facilities, as shown in Fig. 1.7, accounting for 40% of the country's national strategic reserves. The

1.1 Background and Significance

strategic petroleum reserves in France include crude oil, gasoline for vehicles, aviation kerosene, diesel, household fuel, lighting kerosene, jet engine fuel, heavy oil, liquefied petroleum gas, and more. Notably, the Lisieux region stores 1.2 million cubic meters of refined petroleum products, the Marolles region stores 14 million cubic meters of light oil, the Noeux-les-Mines region stores 3 million cubic meters of refined petroleum products, and the Etrez region stores 120,000 cubic meters of refined petroleum products, among others. France's first salt cavern gas storage facility was established in Tersanne in 1968. Subsequently, Etrez, Manosque, and Hauteriver salt cavern gas storage facilities were also built, with a total gas storage capacity of 3 billion cubic meters and a working gas capacity of 1.155 billion cubic meters.

In conclusion, the construction of compressed air energy storage (CAES) power stations using salt caverns offers multiple advantages. It not only enhances the efficiency of clean energy utilization, facilitates peak shaving and load balancing to address the supply–demand disparity in the power grid but also enables the resourceful utilization of abandoned salt mine cavities, meeting the requirements

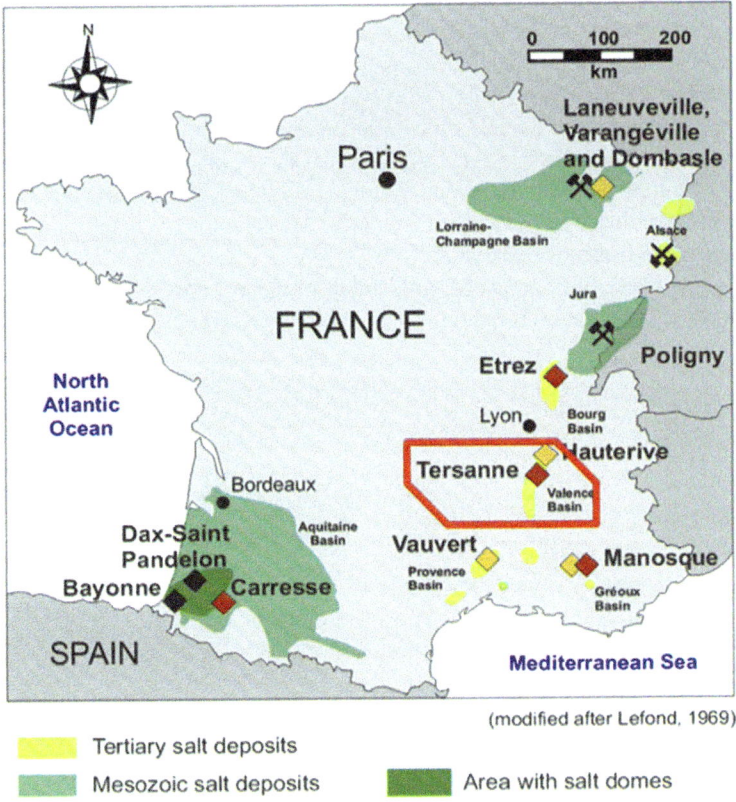

Fig. 1.7 Distribution map of salt mines in France

for sustainable green mining development. Furthermore, it contributes to achieving global goals of carbon peaking and carbon neutrality. One of the key factors in ensuring the stable operation of salt cavern CAES power stations is to guarantee the long-term stability of the salt caverns that store high-pressure air [74].

1.1.3 Significance of the Study of Creep and Fatigue Mechanical Properties of Rock Salts

Salt caverns are subjected to very different stress states in the surrounding rock of the cavity due to their different uses. When used as storage for wastes such as alkali slag and nuclear waste, there is little need to consider the effect of pumping on the stress fluctuations in the surrounding rock of salt cavern reservoirs [64, 75, 76]. Considering the chemical corrosion properties of alkali slag and the continuous exothermic properties of nuclear waste, more research efforts have been focused on the mechanical properties of rock salts subjected to chemical-stress coupling (e.g., alkali slag) [77] and temperature-stress coupling (e.g., nuclear waste) [78, 79]. When used to store oil and natural gas, the salt cavern storage enclosure is more subjected to the creep action of constant stress as the peaking cycle may often be measured in quarters or even years [81, 82]. However, CAES operate differently from all of the above, operating more frequently, even peaking as often as twice a day. Frequent gas injection and exhaust of CAES leads to the salt cavern storage surrounding rock under cyclic loading, which is prone to fatigue. Figure 1.8 shows a typical pressure change cycle of the salt cavern of the Huntorf CAES plant in Germany, and one cycle is about several hours or one day [83].

Simultaneously, during the charging and discharging process, due to the need for load balancing, the rock salt remains in a constant pressure state for a certain period.

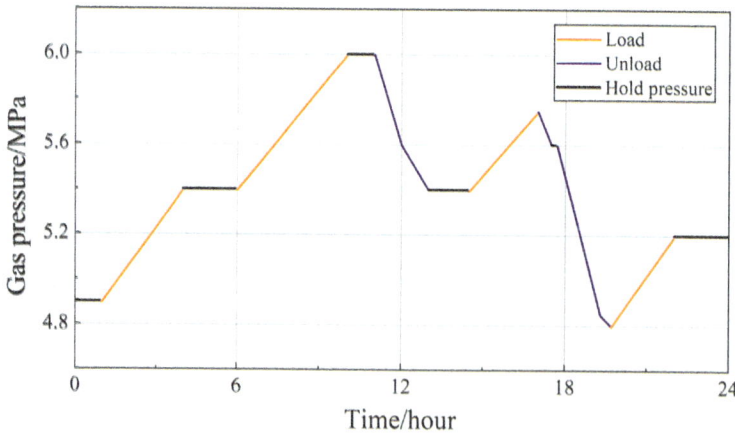

Fig. 1.8 Schematic diagram of the operating gas pressure of the Huntorf CAES plant within 24 h

Additionally, rock salt exhibits good rheological properties, leading to continuous creep deformation during the constant loading period. Consequently, during the operation of a compressed air energy storage (CAES) salt cavern, the surrounding rock is subjected to a dual effect of creep and fatigue. Both creep and fatigue induce irreversible deformation (plastic deformation) [84, 85]. Under the alternating effects of creep and fatigue, the surrounding rock of the salt cavern undergoes processes such as dislocation slip, cumulative damage, and crack propagation, affecting the stability of the salt cavern and potentially leading to instability and failure.

Therefore, studying the interaction between creep and fatigue in rock salt, revealing the evolutionary behavior of creep and fatigue damage, and establishing a constitutive model for creep and fatigue in rock salt are of paramount importance. Investigating the damage, failure, and instability behavior of the CAES salt cavern under different working load conditions can provide scientific recommendations for the global construction of compressed air energy storage power stations utilizing salt caverns. Moreover, this research is of significant reference and guidance for optimizing the operational schemes and improving the economic efficiency of compressed air energy storage power stations.

1.2 State-of-the-Art

Underground salt deposits can be mined by water-soluble mining (e.g., oil cushioning and air cushioning) to form large closed cavities. The extremely low permeability and good rheology of rock salts are the basis for their ability to serve as underground reservoirs for energy. Numerous studies have been carried out on the stability of salt caverns for the storage of oil and gas [86]. This has laid a good foundation for us to study the mechanical properties of the surrounding rock of salt cavern CAES, but considering the special characteristics of the operation mode of CAES plants, which have more frequent peak shifting and a certain length of stabilization period, the fatigue properties of rock salt under cyclic loading, especially the creep–fatigue mechanical properties, are more worthy of our attention. Due to the needs of mining, construction and other industries, researchers have conducted a lot of tests and studies on the fatigue and creep mechanical properties of rocks under cyclic and constant loads [87–90], and achieved good research results. This provides a reference for our further research on the creep–fatigue properties of rock salts.

1.2.1 Fatigue Mechanical Properties of Rocks Under Cyclic Loading

In various geological processes and rock engineering constructions, rocks are often subjected to cyclic loading. For example, geological tectonic processes may lead to crustal deformations influenced by solid tidal forces [92], cyclic seismic impacts forming rock bedding planes [93], and cyclic loading from overlying strata in underground coal mining causing roof subsidence [94]. Due to the cyclic loading, rocks experience fatigue damage, which can significantly impact the stability of underground engineering projects. Therefore, understanding the deformation and failure mechanisms, as well as constitutive models, of rocks under cyclic loading is crucial for accurately assessing the long-term stability of underground rock structures. Researchers studying the fatigue mechanics of rocks have primarily focused on the following aspects.

The deterioration of the mechanical properties of rocks under cyclic loading is referred to as rock fatigue, and the maximum stress level that rocks can withstand without failure under infinite cycles is called rock fatigue strength [95]. The number of cycles causing rock failure under specific loading conditions is known as rock fatigue life. In rock fatigue mechanical testing, researchers mainly study the fatigue characteristics of rocks under different loading conditions such as frequency, waveform, and amplitude, as well as the stress states such as uniaxial, triaxial, tension, compression, and shear, and the environments including dry–wet cycling and freeze–thaw cycling.

Ge et al. [96], concluded that the fatigue deformation of rocks is mainly divided into three stages, namely the initial stage, the isotropic stage and the accelerated stage, and found that the upper limit stress ratio and amplitude are the main factors affecting the fatigue life of rocks by varying the stress ratio cyclic loading and unloading experiments. Guo et al. [97], carried out uniaxial cyclic loading and unloading fatigue experiments on rock salt and found that the axial strain, deformation modulus and fatigue damage process of rock salt can be divided into the same three stages. Ma et al. [98], studied the denaturation, strength and damage characteristics of rock salt under cyclic unloading in triaxial state through different loading waveform parameters and different peripheral pressure experiments of rock salt, and inferred that the upper threshold of deformation damage of rock salt in triaxial cyclic unloading is 80~89%. Zhao et al. [99], carried out dry and wet cyclic tests on mudstone from Sifenshan, Chongqing, and found that dry and wet cycling damaged the pore structure of the mudstone, increased the porosity. With the increase of porosity, the uniaxial compressive strength of the mudstone decreased, and an exponential function between the uniaxial compressive strength and the number of cycles was established. It was also found that the compressive strength of mudstone was related to the number of wet and dry cycles and the initial porosity. Zhu et al. [100], used acoustic emission to study the fatigue characteristics of gypsum plagioclase under cyclic loading, and pointed out that the stress level of cyclic loading has a large influence on the fatigue characteristics. Wang et al. [101], investigated the elastic modulus law of fractured sandstone under cyclic loading by using a combination of laboratory tests and numerical simulation, and obtained that the elastic modulus of most specimens showed the phenomenon of strengthening with the cyclic loading cycles increase. Lu et al. [102], investigated the effect of circumferential pressure on the fatigue characteristics of yellow sandstone, and proposed that the deformation is more suitable than the strength as a criterion for rock damage in the fatigue damage process. Han et al. [103],

1.2 State-of-the-Art

investigated the creep model of rock salt under cyclic loading, proposed a deformation factor based on the test results, and modified the Burgess creep model. Fan et al. [104], carried out a study on the fatigue characteristics of rock salt under discontinuous loads, and found that interval fatigue accelerated the damage accumulation and the fatigue characteristics of the rock salt under discontinuous loads. They found that interval fatigue accelerates damage accumulation and shortens the fatigue life of rock salts, Jiang et al. [105], analyzed the Bauschinger effect and residual stresses as the main causes of this phenomenon from the theoretical point of view. Chen et al. [106], studied the stress state and volume change rule of salt cavity under different gas extraction rates by numerical simulation. Zhou et al. [107], found through the rock salt permeability experiment and Computed Tomography (CT) scanning that the permeability of rock salt changes greatly under the action of seepage and peripheral pressure, and it is consistent with the change of porosity.

In terms of fatigue damage models describing the nonlinear mechanical properties of rocks under cyclic loading, researchers have mainly proposed the following types of constitutive models: classical elastic–plastic constitutive models, constitutive models based on internal variables and fatigue damage models based on energy dissipation [108, 109].

By comparing the generalized Kelvin-Voigt creep constitutive model under low-frequency cyclic/static loading, Huang et al. [110], found that the trends of the two models over time are consistent. The deceleration and stabilization phases of the cyclic loading model are smaller than those of the static loading under the condition of equal upper limit stresses. Drawing on the theory of rock creep, Guo et al. [111]. Defined three fatigue constitutive model elements of elasticity, viscosity, and plasticity, gave the accelerated fatigue parameter n, and established a nonlinear viscoelastic-plastic fatigue constitutive model considering the decelerating, stabilizing, and accelerating phases. The fatigue acceleration parameter n was found to be inversely related to the uniaxial compressive strength of the rock. Cerfontaine et al. [112], proposed an intrinsic model suitable for describing the fatigue damage of brittle rocks based on the theory of cumulative plastic strain and the assumption of reducing the cohesion of the rock material by cyclic loading. And the model was validated using three kinds of rocks: marble, sandstone and granite. The results show that the model can reproduce the change trend of fatigue test and can effectively predict the fatigue life. Zhang et al. [113], investigated the microstructural changes of rock salt under cyclic loading, and pointed out that the fracture extension inside the rock salt was mainly dominated by inter-crystal fractures, and the number of fractures was positively correlated with the upper limit stress, and the macro porosity inside the rock salt showed a trend of increase in the total porosity with the development of fatigue test. An empirical fatigue model was obtained by solving the S-shaped inverse function and verified. Yang et al. [114], investigated the strength change and fatigue deformation characteristics of coal rock under cyclic loading, and the results showed that the stress threshold for fatigue damage occurred in coal rock was lower compared with hard rock. The damage of coal rock can be predicted based on the change of unloaded elastic modulus of radial deformation of coal rock, and the development of coal rock damage variables with cyclic loading was fitted according to the theory of continuum damage mechanics. Dai et al. [115], based on the assumption of Lemaitre's strain equivalence, combined with Weibull's statistical damage model of microscopic defects and the fracture mechanics model of macroscopic joints, described the deformation of jointed rock under cyclic loading and the degradation behavior of strength properties of nodular rock under cyclic loading. An intrinsic fatigue damage model based on residual strain is also proposed. By defining new fatigue damage variables, Li et al. [116], based on the high circumferential fatigue damage model, the hardening parameter, which has a power law relationship with the stress amplitude and the number of cycles, was introduced to derive the evolution equation of low circumferential fatigue damage in rocks, and the model was verified using the measured fatigue data of sandstone. The model is found to be able to fit the evolution curve of the low-cycle fatigue damage of rocks better.

The aforementioned research on the fatigue mechanics and damage evolution of rocks has laid the groundwork for further accurately assessing the stability of the surrounding rock of salt cavern storage under gas injection and production conditions.

1.2.2 Creep Mechanical Properties of Rocks Under Constant Loads

The deformation of rock under constant load increases with time, called creep, is one of the most important characteristics of rock [117]. In the process of underground geotechnical engineering, especially deep geotechnical engineering, construction and operation, the deformation of surrounding rock due to creep and the destabilization of the structure is one of the main reasons affecting the safety of underground engineering, and the study of the creep mechanical properties of rock is of great significance to ensure the long-term stability of underground building structures. Since the 1920s, researchers have carried out a large number of creep tests different conditions, summarized in many creep deformation laws [118–121].

Li et al. [123], carried out long-time graded creep tests on serpentine marble and found that the steady-state creep rate and creep strain showed an increasing trend with the increase of stress level and time. By comparing the inflection points of isochronous stress–strain curves, the long-term creep strength of serpentine marble was determined to be around 35 MPa. Hu et al. [124], simulated the uniaxial creep mechanical properties of non-homogeneous granite by using PFC^{2D} particle flow numerical simulation software, by setting different grain diameters, and combining the equivalent crystalline model and stress corrosion model. The results show that the homogeneity of the rock shows a positive correlation with the uniaxial compressive strength and long-term creep strength, and the non-homogeneous factor shows a negative correlation with the damage time of creep failure under a given stress level, and the microstructure of the rock controls the damage form of rock creep. Zhou et al. [125], studied the creep deformation law of gas coal under high-temperature triaxial stress, pointing out that when the temperature is low (200 °C), during the

1.2 State-of-the-Art

100-h creep experiment, the gas coal only shows the two-stage characteristics of creep. While at higher temperature (400 °C), the gas coal entered the accelerated creep stage after only 6 h. It is confirmed that the increase of temperature determines the creep process of gas-coal, and combined with the change rule of porosity, it is judged that 300 °C is the critical temperature at which the creep state of gas-coal is changed. Zhang et al. [126], gave the creep damage characteristics of mudstone subjected to dynamic/static load, and found that the larger the axial static load, the greater the influence of disturbance amplitude and frequency on the creep damage of mudstone, and the creep time is negatively correlated with the amplitude and frequency. The addition of dynamic loading led to a decrease in the strength of the particles within the rock, weakened the bonding effect between particles, and accelerated the expansion of intergranular and intragranular cracks. Zafar et al. [127], used Barre granite containing double defects to investigate the creep and relaxation experiments of brittle rocks at multiple stress levels, and the results showed that when the stress level exceeded the crack damage threshold, the failure time of the creep is much smaller than that of the loosely failure time. Creep produces higher cumulative acoustic emission events and event rates than relaxation. The b-value of AE derived by the maximum likelihood method can be used as a precursor to macroscopic damage in rock specimens. The above studies are an important contribution to the understanding of the creep properties of rocks, but are mainly aimed at brittle rocks with high hardness.

Rock salt, as a typical soft rock, has strong rheological properties and can produce large deformation at low stress [128]. Researchers have also carried out a large number of salt-rock creep tests. Yang et al. [129], used digital image correlation (DIC) to study the microstructural changes and damage evolution in the creep process of salt-rock. Ma et al. [130], investigated the effect of gypsum content on the creep rate of salt-gypsum strata, and found that the steady-state creep rate of high gypsum salt-rock strata was lower than that of high salt-salt-salt-rock strata under the same temperature and stress conditions. Lance et al. [131], investigated the effect of cyclic loading and unloading on the spreading characteristics and creep response of the rock salt of Avery Island, U.S.A., and found that cyclic loading and unloading did not have much impact on the creep. Li et al. [132], investigated the effect of temperature on the creep damage evolution process of rock salts using the discrete element method. Luangthip et al. [133], investigated the content of carnallite in the Maha Sarakham rock salt, Thailand, and its correlation with creep properties. Hamami et al. [134], investigated the strain hardening properties of the rock salt during the creep process by using triaxial multi-stage creep tests.

Establishing a model that can accurately predict the long-term creep of rock has been key to evaluate whether the underground geotechnical engineering can operate stably for a long time. Existing rock creep models are mainly categorized into the following three types: (1) creep empirical models [135], (2) component combination models [136], (3) fracture damage models [137].

Empirical rock creep models are fitted based on measured data from rock creep tests to construct correlation functions of stress, strain and time. Common types of functions for empirical modeling of creep include power law, logarithmic, exponential, and hybrid equations of the three. The empirical equation for rock creep can generally be expressed in the following three-stage form,

$$\varepsilon(t) = \varepsilon_e + \varepsilon_1(t) + Vt + \varepsilon_3(t) \tag{1.1}$$

where $\varepsilon(t)$ is the total creep strain, ε_e is the instantaneous elastic strain produced by the instantaneous loading of stress, $\varepsilon_1(t)$ is the creep strain in the decelerated creep stage, Vt is the creep strain in the stable deformation stage, where V is the steady-state creep rate, and $\varepsilon_3(t)$ is the creep strain in the accelerated creep stage. For different rocks or different experimental conditions, the corresponding empirical creep model can be obtained directly by fitting experimental data. Wu et al. [138], found that the creep of coal rock conforms to the logarithmic empirical formula by studying the creep test of coal rock, and obtained the parameters of the empirical formula for the creep of coal rock under different stress conditions. Ding et al. [139], established the creep equation of soft and weak interlayer by studying the creep characteristics of soft and weak interlayer in open slope. Zhang et al. [140], carried out triaxial creep test on mudstone, and obtained the nonlinear creep equation of mudstone through the creep empirical model. The model is easy to understand, has high fitting accuracy, and can guide the relative rock engineering. However, the empirical model can only reflect the deformation law of rock creep under loading paths corresponding to the test conditions. Taking the deceleration creep as an example, the typical deceleration creep equation is

$$\varepsilon_1(t) = At^n, 0 < n < 1 \tag{1.2}$$

$$\varepsilon_1(t) = A \ln t \tag{1.3}$$

$$\varepsilon_1(t) = A \ln(1 + at) \tag{1.4}$$

$$\varepsilon_1(t) = A\left[(1 + at)^n - 1\right] \tag{1.5}$$

where A, n and a are constants with appropriate dimensions. These functions can usually fit the strain data in some limited time period, but not in the entire test time, and the physical significance of the parameters in the model is usually not clear. It is therefore difficult to reflect the intrinsic mechanism of the rock creep damage, and cannot be promoted and applied in other geotechnical engineering, adaptability is poor.

Element combination model is the elastic unit (Hooker's body), viscous unit (Newton's body), plastic unit (St. Vernand's body) through the series–parallel combination of creep model. Currently more common element combination model is given by Maxwell model,

1.2 State-of-the-Art

$$\varepsilon = (\sigma_o/k) + (\sigma_o t/n) \tag{1.6}$$

Or Kelvin model

$$\varepsilon = (\sigma_o/k)\left[1 - e^{-kt/n}\right] \tag{1.7}$$

where k is the stiffness factor, η is the viscosity factor, and σ_o is the instantaneous loading stress.

Models such as Burgers, viscoplastic and Nishihara: some generalized creep models consisting of several identical models in series or parallel (e.g., generalized Maxwell model and generalized Kelvin model) are also used when needed. Classical component combination models are widely used in the mechanical analysis of rock underground engineering due to the clear physical meaning of their parameters and the ability to visually characterize the complex mechanical properties of rocks [141–143]. Compared with the empirical model, the unit combination model can more accurately respond to the elastic, viscous and plastic characteristics of rock-like materials. Moreover, the combination can be closer to the actual situation of rock creep by adjusting the number of units. Researchers have carried out a lot of research. Jiang et al. [145], by connecting the strain-triggered inertial element in series with the classical Nishihara model, proposed an extended Nishihara model, which can fit the triaxial creep test curve of sandstone very well. The results show that the smaller the value of inertia mass, the faster the axial strain increases in the creep acceleration stage, the shorter the time to failure, and the failure time of sandstone creep is highly dependent on the applied stress level. Li et al. [146], with the help of the Burgers creep model, through the introduction of fractional-order Abel viscous unit and aging factor, established a new three-dimensional creep model that can respond to the deformation of soft rock under the water–rock interaction with time, and achieved good results. A new three-dimensional creep model was developed by introducing a fractional order Abel viscous unit and a time-dependent factor that can reflect the deformation of soft rock with time under water–rock interaction, and good fitting results were achieved as well. A viscoplastic model that reproduces the creep behavior and inelastic deformation of rock during the loading–unloading cycle was developed by Haghighat et al. [147], Maximization of dissipated energy during plastic flow using Perzyna type of viscous deformation was combined with an improved Cam-clay plastic deformation model. This plastic flow model is of the correlation type, where the viscous deformation is proportional to the ratio of the driving stress to the material viscosity. However, in the models, the constitutive parameters of the unit cells, such as elastic modulus, viscosity coefficient, and yield strength, are often treated as constants. Even with the presence of time variables, deformation remains linear under given time conditions. As a result, it can only reflect the linear changes in the rock creep process. However, most rocks in the natural environment are heterogeneous materials, and their changes are not linear. Especially in the accelerated creep stage, the nonlinear characteristics of rocks become more pronounced, leading to poor consistency between the theoretical curves of traditional component combination models and experimental data [148, 149]. In order to reflect the nonlinear relationship between stress and strain in the creep process, many scholars began to study the nonlinear creep model of rocks, and new creep models were proposed by considering the creep parameters of rocks as time-dependent variables. Hou et al. [151] proposed a four-component model reflecting the creep behaviors of rocks with different initial damage value on the basis of the results of the creep tests of sandstones. Cheng et al. [152], considered the viscoelastic-plastic characteristics and damage evolution of rocks. Building upon the improved Nishihara model, they proposed a nonlinear rock creep model that incorporates damage evolution to predict rock creep behavior. This model introduces damping elements and nonlinear viscoplastic bodies into the classical Nishihara model to simulate the accelerated creep stage of soft rocks. Zhou et al. [153], proposed a fractional-order creep model for rock salt by replacing the viscous unit in the Nishihara model with a variable-order Abel damper based on the theory of damage mechanics of continuous media and the theory of fractional-order derivatives. Wang et al. [154], considered the strength and elastic modulus of the rock as a function of the creep time, and proposed a nonlinear, non-smooth, plastic-viscous eigen-model to reflect the nonlinear creep behavior.

In recent years, with the continuous development of rock damage mechanics and fracture mechanics, many scholars try to introduce new theories of damage mechanics and fracture mechanics to establish a rock creep model based on fracture damage theory [155, 156]. Zhou et al. [157], proposed a nonlinear creep model considering damage on the basis of improved Nishihara model, Chen et al. [158], based on the relationship between damage mechanics and elastic modulus, established an intrinsic model of damage creep that accurately reflects the creep characteristics of rock under acid corrosion and true triaxial stress, and found that the model not only accurately describes the creep properties of rock in the transient elastic strain stage and isokinetic creep stage, but also describe the nonlinear characteristics of the creep curves in the decay creep stage and accelerated creep stage. Hoxha et al. [159], analyzed the effect of relative humidity on the creep of gypsum, and introduced damage variables to establish a damage model for gypsum creep, which better describes the accelerated creep stage.

By applying damage theory to establish damage models, it is possible to provide a more comprehensive reflection of the creep and damage behavior of rocks. In the analysis of rock creep, rock damage can lead to inelastic flow and creep. Damage theory can be applied to stage-wise damage accumulation, ultimately leading to accelerated creep failure. Fracture mechanics is a discipline that studies material crack strength and crack propagation. Due to the discontinuity and heterogeneity of rocks, using fracture mechanics methods to study the rheological mechanisms of rocks is of significant importance. The fracture creep models of rocks have also been extensively studied and have achieved numerous achievements [160, 161].

1.2 State-of-the-Art

1.2.3 Advancements in Rock Creep–Fatigue Mechanics Research

The research on rock creep and fatigue, as described above, has played a crucial role in accurately assessing the damage evolution of the surrounding rock in salt caverns used for compressed air energy storage (CAES) power plants. However, considering the actual operational mode of CAES power plants during gas injection and withdrawal cycles, as shown in Fig. 1.8, there are time intervals where the surrounding rock is subjected to creep–fatigue alternating actions. Therefore, studying the interaction between rock creep and fatigue, and establishing a creep–fatigue constitutive model for rock under loading and unloading, can provide a more precise evaluation of the stability of the salt caverns used as storage reservoirs. This research holds significant importance in ensuring the safe operation of compressed air energy storage power plants [15].

The concept of creep–fatigue interaction was originally proposed by researchers studying the fatigue mechanics of metals at high temperatures. It was found that metals used at high temperatures often experience an interaction between creep and fatigue under low-frequency cyclic loading conditions, leading to premature failure compared to the theoretically calculated fatigue life. In recent years, several researchers have investigated the mechanical properties of various alloys under the interaction of fatigue and creep–fatigue. Their studies have indicated that stress levels, dwell times, temperature variations, and test environments directly influence the effects of the interaction between fatigue and creep in metals [162–166]. For instance, Asada et al. [168], developed a model based on damage mechanics to describe the creep–fatigue interaction behavior of 9Cr-1Mo-V-Nb steel under constant dwell time conditions. They pointed out that fatigue and creep are comprised of component damage during each cycle, and are given by the product of the number and size of components. The creep–fatigue interaction behavior is considered to be caused by the early growth of creep component damage size due to fatigue loading during each cycle, resulting in early failure. Wang et al. [169], systematically studied the creep–fatigue interaction behavior of a nickel-based single crystal superalloy under various strain (tensile) dwell conditions at 900 °C using an in-situ SEM platform. They found that the initial cyclic life rapidly decreases with increasing dwell time, followed by a slight increase in cyclic life for dwell times exceeding 60 s. This phenomenon was attributed to the accumulation of non-elastic strains under different dwell conditions. Non-elastic strain accumulates more under dwell fatigue conditions than pure fatigue loading, and increases with dwell time. However, after a dwell time of 60 s, it decreases again due to the formation of a more stable network structure at longer dwell times. Goyal et al. [170], developed a model to represent the creep–fatigue damage mechanism resulting from the combined effects of fatigue, creep, oxidation, and corrosion, which depends on various loading conditions. They pointed out that under creep and oxidation conditions, cracks mainly propagate along grain boundaries starting from the notch surface, while under fatigue-oxidation conditions, cracks propagate trans granularly. Numerous studies have indicated that as stress dwell time and level increase, the fatigue life of materials significantly decreases, demonstrating the significant interaction between creep and fatigue. Therefore, it is necessary to study the mechanical properties of other materials under creep–fatigue loading conditions.

As some materials will often be affected by creep–fatigue loading, some researchers have also carried out experimental studies on the mechanical properties of these materials under creep–fatigue interaction, Zhu et al. [171], studied the fatigue damage characteristics of asphalt mixtures under cyclic loading, and proposed a damage evolution equation for asphalt considering the creep–fatigue damage effect. The study shows that the fatigue damage of asphalt materials is due to the joint damage effect of creep and fatigue. Qi et al. [172], analyzed the creep mechanical properties of saturated soft clay under cyclic loading, established an empirical equation of cumulative plastic strain that can be applied in practical engineering and imported into Abaqus finite element numerical simulation software for secondary development. It was found that the deformation of saturated soft clay increased greatly after considering the creep effect. Zhou et al. [173], carried out the creep fatigue test of plastic wood composite flooring, and the study showed that the strain curve of creep–fatigue of plastic wood flooring is similar to the pure creep curve, which is divided into three stages. With the increase of creep load percentage, the life of plastic wood flooring decreases, creep, fatigue interaction will reduce the strength of plastic wood flooring. Sain et al. [174], investigated the fatigue creep behavior of thermoplastic wood fiber composites, both unmodified and functionally modified. For PVC, PE, and PP-based composite materials, creep is largely influenced by the load level, time, and temperature. Slight elevations in temperature above ambient conditions significantly enhance creep in PVC-wood fiber composites.

In the field of rocks, researchers also found that fatigue and creep are similar in time dimension, that is, the deformation curve can be divided into three stages: decelerated deformation-steady deformation-accelerated deformation [175], and the creep eigen-structure model can also be used to describe the damage evolution process of fatigue in rocks. Wang et al. [176], investigated the mechanical properties of the creep of rock salt under the action of long-term cyclic loading, and pointed out that the upper limit of the stress determined the development trend of the stress–strain curve of the rock salt. It is pointed out that the upper stress limit determines the development trend of the stress–strain curve of rock salt, and the average maximum creep rate in the stable creep stage decreases with the increase of the cycle and increases with the increase of the upper stress limit. Based on Kachanov creep damage theory, Miao et al. [177], established a fatigue rheological damage model of rock under cyclic loading by combining with Burgers model. By substituting the sinusoidal stress function into the constant stress in the rheological differential eigenstructure equation, the one-dimensional and three-dimensional differential-type damage eigenstructure equations of rocks under cyclic loading were derived. The model can accurately describe the fatigue rheological damage development trend of siltstone under cyclic loading. Xu et al. [178], modified the Poyhting-Thomson rheological fatigue model according to the real fatigue process under cyclic loading of rock salt, using the time-dependent power function to describe the strain increment caused by fatigue

1.2 State-of-the-Art

damage, and the results showed that the model can better describe the strain time curve of rock salt. Through the above studies, it can be found that researchers tend to transform the time function of creep into a function related to the number of load actions to characterize the damage of the material, for rocky materials, especially rock salt, scholars have investigated the creep properties of rock salt under cyclic loading, which is regarded as a function related to time. However, the above-mentioned studies mainly involve deriving constitutive equations and fitting data to illustrate the interaction between creep and fatigue. There are relatively few experiments that directly demonstrate the mutual influence between creep and fatigue in rocks [179, 180, 181]. For the rock low-stress creep–fatigue test, researchers have conducted some studies, Cui et al. [182], studied the interval fatigue characteristics of rock salt under uniaxial conditions using acoustic emission, and found that the time interval accelerates the accumulation of damage in rock salt, and this phenomenon becomes more pronounced with longer time intervals. Jiang et al. [183], investigated the interval fatigue characteristics of concrete and proposed that residual stresses generated from non-coherent deformation in concrete can promote the damage of concrete specimens during the interval duration. In comparison, high-stress creep–fatigue tests on rocks are rarely reported, with only a few studies focused on sandstone, rock salt, and other rock types. Wang et al. [184], conducted creep–fatigue loading tests on prefabricated angle-cracked sandstone and found that after creep loading, the deformation modulus and secant modulus of the sandstone significantly decreased, increasing the energy dissipation of the rock and accelerating sample damage. Shi et al. [185], used acoustic emission to study the influence of creep on the fatigue mechanics of red sandstone specimens with prefabricated cracks under different stress levels. They found that applying short-term, low-stress creep loads can reduce the increase of fatigue deformation parameters, slow down the decay of elastic parameters, and delay the occurrence of Felicity effect (The Felicity effect in Physics, is an effect observed during acoustic emission in a structure undergoing repeated mechanical loading.) in the rock mass, with a slower Felicity decay. Ma et al. [186], investigated the creep–fatigue mechanics of rock salt under acoustic emission monitoring and found that under the combined action of creep and fatigue loads, the cyclic life of rock salt significantly decreased compared to what is observed in pure fatigue or pure creep tests, indicating a clear interaction between fatigue damage accumulation and creep damage accumulation. With the increase of creep loading time, the creep life of rock salt showed a linear increase, while the fatigue life exhibited an exponential decrease. Some researchers also used equipment such as computer tomography (CT) and low-field nuclear magnetic resonance (NMR) to study the internal microdamage evolution of rocks under creep–fatigue loading [187]. The above-mentioned research on material creep–fatigue mechanics can provide references for analyzing the creep–fatigue mechanics characteristics of rock salt.

1.3 Research Content

Against the backdrop of carbon peak and carbon neutrality, global sustainable energy development has ushered in new opportunities. This paper takes the construction and operation of salt cavern compressed air energy storage (CAES) power plants as the background and considers the unique peaking demand of salt cavern CAES power plants. The surrounding rock of salt cavern storage will be subjected to the dual effects of creep and fatigue. Through laboratory experiments, theoretical analysis, and model studies, we designed uniaxial and triaxial compressive strength tests of rock salt under different confining pressures, creep tests under different stress levels, fatigue tests under different loading and unloading rates, and creep–fatigue tests of rock salt with different low-stress interval durations, creep–fatigue tests with different high-stress interval times, creep–fatigue tests under different confining pressures, and long-term creep–fatigue tests. With the help of observation equipment such as acoustic emission, we analyzed in detail the macroscopic crack distribution and microscopic damage evolution of rock salt under creep–fatigue alternation and established a rock salt creep–fatigue mechanical constitutive model that considers the interaction between creep and fatigue. The specific research contents are as follows:

(1) **Basic mechanical properties tests of rock salt**

Basic mechanical properties are the foundation for creep and fatigue tests. Scanning electron microscopy (SEM) and X-ray diffraction (XRD) were used to identify the chemical composition and structure of the salt. Triaxial and uniaxial compression tests were conducted on rock salt, and the test data were used to calculate uniaxial compressive strength (UCS), elastic modulus, Poisson's ratio, friction angle, and cohesion. The dilatancy effect, which leads to volume expansion, increased permeability/porosity, and crack growth, can induce mechanical instability in engineering. The general characteristics of salt dilatancy were summarized, and the compression-dilatancy transition point was identified.

(2) **Conventional mechanical properties of creep and fatigue in rock salt**

Creep and fatigue tests were conducted on rock salt under different stress levels and loading rates. The effects of different loading methods, stress levels, and loading/unloading rates on overall deformation, residual strain, creep deformation/rate of rock salt were analyzed. It was proposed that the deformation of rock salt during the creep and fatigue processes can be divided into time-dependent deformation (creep plastic deformation) and time-independent deformation (loading plastic deformation).

(3) **Discontinuous Fatigue Mechanical Properties of rock salt**

Discontinuous fatigue tests were conducted on rock salt under different stress levels and varying interval times. The effects of different loading methods, stress levels, and low-stress interval times on the fatigue life, residual strain, and Poisson's ratio of rock salt were analyzed. It was found that the presence of low-stress intervals accelerates the damage to rock salt.

(4) **Creep–fatigue mechanical characterization of rock salt under unidirectional stresses**

Uniaxial creep–fatigue tests of rock salt with different high stress plateaus were carried out to analyze the damage evolution of rock salt creep–fatigue under different high stress intervals, and to explain the effect of rock salt creep–fatigue interaction on the residual strain and creep strain of rock salt from a microscopic point of view.

(5) **Creep–fatigue mechanical characterization of rock salt under three-way stresses**

The triaxial creep–fatigue test of rock salt under different peripheral pressures was carried out to study the influence of different peripheral pressures on the creep–fatigue stress–strain curves of rock salt, analyze the influence of peripheral pressure densification effect on the creep–fatigue interactions of rock salt, and reveal the influence of the construction depth of the pressurized-gas storage plant on the peripheral rock of the salt cavern storage.

(6) **Multilevel amplitude creep–fatigue mechanical characterization of rock salt under acoustic emission monitoring**

Uniaxial/triaxial creep–fatigue tests of rock salt under different stress levels were carried out to study the effects of different stress levels on the creep–fatigue mechanical properties of rock salt. With the help of acoustic emission monitoring equipment, the effect of multi-level amplitude creep/fatigue loading on the damage evolution of rock salt was analyzed.

(7) **Long-time creep–fatigue mechanical characterization of rock salt**

Based on the real peaking frequency of the compressed air energy storage (CAES) power plant, long-term creep–fatigue tests on rock salt were conducted. On a relatively long-time scale, the influence of different operating pressure upper limits and cycles on the creep–fatigue mechanical properties of rock salt was studied, and the effect of stress state variations on rock salt creep–fatigue deformation was analyzed in detail.

(8) **Creep–fatigue constitutive modeling of rock salt considering creep–fatigue interaction**

Based on the existing research data, a new creep–fatigue constitutive model for rock salt was developed by incorporating a state variable that characterizes the degree of rock hardening into the Norton creep model. This model takes into account the interaction between creep and fatigue in rock salt. The model was validated using test data from rock salt creep–fatigue mechanical experiments.

References

1. Leon Hermanson, Smith Doug, Seabrook Melissa, et al. WMO global annual to decadal climate update: A prediction for 2021–25[J]. Bulletin of the American Meteorological Society, 2022, 103(4): E1117-E1129.

2. Michael-F Wehner, Reed Kevin-A. Operational extreme weather event attribution can quantify climate change loss and damages[J]. PLOS Climate, 2022, 1(2): e13.
3. Rebecca Lindsey. Climate change: Global sea level[J]. Available online: Climate. gov (accessed on 14 August 2020), 2021.
4. Piers-M Forster, Smith Christopher-J, Walsh Tristram, et al. Indicators of Global Climate Change 2022: annual update of large-scale indicators of the state of the climate system and human influence[J]. Earth System Science Data, 2023, 15(6): 2295-2327.
5. Cara-A Horowitz. Paris agreement[J]. International Legal Materials, 2016, 55(4): 740-755.
6. Carl-Friedrich Schleussner, Rogelj Joeri, Schaeffer Michiel, et al. Science and policy characteristics of the Paris Agreement temperature goal[J]. Nature Climate Change, 2016, 6(9): 827-835.
7. John Pezzey. Sustainable development concepts[J]. World, 1992, 1(1): 45.
8. John McCormick. The origins of the world conservation strategy[J]. Environmental Review, 1986, 10(3): 177–187.
9. Gro-Harlem Brundtland. What is sustainable development[J]. Our common future, 1987, 8(9).
10. Glen-P Peters, Andrew Robbie-M, Canadell Josep-G, et al. Key indicators to track current progress and future ambition of the Paris Agreement[J]. Nature Climate Change, 2017, 7(2): 118-122.
11. Spencer Dale. BP statistical review of world energy[J]. BP Plc: London, UK, 2021, 14–16.
12. Dolf Gielen, Gorini Ricardo, Wagner Nicholas, et al. Global energy transformation: a roadmap to 2050[J]. 2019.
13. Elisa Asmelash, Prakash Gayathri, Gorini Ricardo, et al. Role of IRENA for global transition to 100% renewable energy[J]. Accelerating the transition to a 100% renewable energy era, 2020, 51–71.
14. Dahai Zhang, JiaqiWang, Yonggang Lin, et al. Present situation and future prospect of renewable energy in China[J]. Renewable and Sustainable Energy Reviews, 2017, 76865–871.
15. Zongze Li, Zhenyu Yang, JinYang Fan, et al. Fatigue mechanical properties of salt rocks under high stress plateaus: the interaction between creep and fatigue[J]. Rock Mechanics and Rock Engineering, 2022, 55(11): 6627-6642.
16. Wei Liu, Zhixin Zhang, Jie Chen, et al. Feasibility evaluation of large-scale underground hydrogen storage in bedded salt rocks of China: A case study in Jiangsu province[J]. Energy, 2020, 198117348.
17. JinYang Fan, Heping Xie, Jie Chen, et al. Preliminary feasibility analysis of a hybrid pumped-hydro energy storage system using abandoned coal mine goafs[J]. Applied Energy, 2020, 258114007.
18. Ning Zhang, Lu Xi, McElroy Michael-B, et al. Reducing curtailment of wind electricity in China by employing electric boilers for heat and pumped hydro for energy storage[J]. Applied energy, 2016, 184987–994.
19. Mengye Zhu, Qi Ye, Belis David, et al. The China wind paradox: the role of state-owned enterprises in wind power investment versus wind curtailment[J]. Energy Policy, 2019, 127200–212.
20. Xiaowei Ma, Zhiren Zhang, Hewen Bai, et al. A Mid/Long-Term Optimization Model of Power System Considering Cross-Regional Power Trade and Renewable Energy Absorption Interval[J]. Energies, 2022, 15(10): 3594.
21. Jin Xiao, Guohao Li, Xie Ling, et al. Decarbonizing China's power sector by 2030 with consideration of technological progress and cross-regional power transmission[J]. Energy Policy, 2021, 150112150.
22. Xu Peng, Guangle Gao, Gaoge Hu, et al. Research on inter-regional renewable energy accommodation assessment method based on time series production simulation[A]//IEEE, 2019: 2031–2036.
23. Amit-Kumar Rohit, Devi Ksh-Priyalakshmi, Rangnekar Saroj. An overview of energy storage and its importance in Indian renewable energy sector: Part I–Technologies and Comparison[J]. Journal of Energy Storage, 2017, 1310–23.
24. John-P Barton, Infield David-G. Energy storage and its use with intermittent renewable energy[J]. IEEE transactions on energy conversion, 2004, 19(2): 441-448.

References

25. S-Ould Amrouche, Rekioua Djamila, Rekioua Toufik, et al. Overview of energy storage in renewable energy systems[J]. International journal of hydrogen energy, 2016, 41(45): 20914–20927.
26. A-G Olabi. Renewable energy and energy storage systems. Elsevier, 2017: 1–6.
27. Lei Zhang, Tianshu Zhou, Yongxia Zhao, et al. Research on Electricity Cost Changes of Consumer from Developing Renewable Energy during the 14th Five-Year Plan Period — Case Studies of Ningxia Power Grid[J]. Theory & Practice, 2022, (04): 102-105.
28. Weihe Huang, Jingkuan Han, YushengWang, et al. Strategies and Countermeasures for Ensuring Energy Security in China[J]. Strategic Study of CAE, 2021, 23(1): 112–117.
29. Mathew Aneke, Wang Meihong. Energy storage technologies and real life applications–A state of the art review[J]. Applied Energy, 2016, 179350–377.
30. TMI Mahlia, Saktisahdan T-J, Jannifar A, et al. A review of available methods and development on energy storage; technology update[J]. Renewable and sustainable energy reviews, 2014, 33532–545.
31. Md-Mustafizur Rahman, Oni Abayomi-Olufemi, Gemechu Eskinder, et al. Assessment of energy storage technologies: A review[J]. Energy Conversion and Management, 2020, 223113295.
32. K-C Divya, Østergaard Jacob. Battery energy storage technology for power systems—An overview[J]. Electric power systems research, 2009, 79(4): 511–520.
33. Yury Gogotsi. Energy storage wrapped up[J]. Nature, 2014, 509(7502): 568-569.
34. WeilongWang, Guo Shaopeng, Li Hailong, et al. Experimental study on the direct/indirect contact energy storage container in mobilized thermal energy system (M-TES)[J]. Applied energy, 2014, 119181–189.
35. Venkata-Suresh Vulusala G, Madichetty Sreedhar. Application of superconducting magnetic energy storage in electrical power and energy systems: a review[J]. International Journal of Energy Research, 2018, 42(2): 358–368.
36. SD-Gamini Jayasinghe, Vilathgamuwa D-Mahinda, Madawala Udaya-K. Direct integration of battery energy storage systems in distributed power generation[J]. IEEE Transactions on Energy Conversion, 2011, 26(2): 677–685.
37. John-B Goodenough. Electrochemical energy storage in a sustainable modern society[J]. Energy & Environmental Science, 2014, 7(1): 14-18.
38. Kun Zhang, Bo Peng, Jiaojiao Guo, et al. Application Status and Prospective Analysis of Chemical Energy Storage Technology in Large-scale Energy Storage Field[J]. Power Capacitor & Reactive Power Compensation, 2016, 37(2): 54–59, 66.
39. Cong Ding, Huamin Zhang, Xianfeng Li, et al. Vanadium flow battery for energy storage: prospects and challenges[J]. The journal of physical chemistry letters, 2013, 4(8): 1281-1294.
40. KangliWang, Kai Jiang, Chung Brice, et al. Lithium–antimony–lead liquid metal battery for grid-level energy storage[J]. Nature, 2014, 514(7522): 348–350.
41. Montaser Mahmoud, Ramadan Mohamad, Olabi Abdul-Ghani, et al. A review of mechanical energy storage systems combined with wind and solar applications[J]. Energy Conversion and Management, 2020, 210112670.
42. Mustafa-E Amiryar, Pullen Keith-R. A review of flywheel energy storage system technologies and their applications[J]. Applied Sciences, 2017, 7(3): 286.
43. Jinyang Fan, Wei Liu, Deyi Jiang, et al. Thermodynamic and applicability analysis of a hybrid CAES system using abandoned coal mine in China[J]. Energy, 2018, 15731–44.
44. Deyi Jiang, Shao Chen, Wenhao Liu, et al. Underground hydro-pumped energy storage using coal mine Goafs: system performance analysis and a case study for China[J]. Frontiers in Earth Science, 2021, 9947.
45. Ahmad Arabkoohsar. Chapter Eight - Conclusion. Academic Press, 2021: 177–188.
46. Mike McWilliams. 6.08 - Pumped Storage Hydropower. Oxford: Elsevier, 2022: 147–175.
47. Ryan Wiser, Bolinger Mark, Hoen Ben, et al. Land-based wind market report: 2022 edition. Lawrence Berkeley National Lab.(LBNL), Berkeley, CA (United States), 2022.
48. Alessandro Tallini, Vallati Andrea, Cedola Luca. Applications of micro-CAES systems: energy and economic analysis[J]. Energy Procedia, 2015, 82797–804.

49. Hussein Ibrahim, Younès Rafic, Ilinca Adrian, et al. Study and design of a hybrid wind–diesel-compressed air energy storage system for remote areas[J]. Applied Energy, 2010, 87(5): 1749-1762.
50. Rui Li, Laijun Chen, Tiejiang Yuan, et al. Optimal dispatch of zero-carbon-emission micro Energy Internet integrated with non-supplementary fired compressed air energy storage system[J]. Journal of Modern Power Systems and Clean Energy, 2016, 4(4): 566-580.
51. Qian Zhou, Du Dongmei, Lu Chang, et al. A review of thermal energy storage in compressed air energy storage system[J]. Energy, 2019, 188 115993.
52. Elaheh Bazdar, Sameti Mohammad, Nasiri Fuzhan, et al. Compressed air energy storage in integrated energy systems: A review[J]. Renewable and Sustainable Energy Reviews, 2022, 167 112701.
53. Jie Chen, Song Ren, Chunhe Yang, et al. Self-healing characteristics of damaged rock salt under different healing conditions[J]. Materials, 2013, 6(8): 3438-3450.
54. Mengsu Hu, Steefel Carl-I, Rutqvist Jonny, et al. Microscale THMC Modeling of Pressure Solution in Salt Rock: Impacts of Geometry and Temperature[J]. Rock Mechanics and Rock Engineering, 2022, 1–19.
55. A Mortazavi, Nasab H. Analysis of the behavior of large underground oil storage caverns in salt rock[J]. International Journal for Numerical and Analytical Methods in Geomechanics, 2017, 41(4): 602–624.
56. Janos-L Urai, Spiers Christopher-J, Zwart Hendrik-J, et al. Weakening of rock salt by water during long-term creep[J]. Nature, 1986, 324(6097): 554-557.
57. Dilara-Gulcin Caglayan, Weber Nikolaus, Heinrichs Heidi-U, et al. Technical potential of salt caverns for hydrogen storage in Europe[J]. International Journal of Hydrogen Energy, 2020, 45(11): 6793-6805.
58. Zongze Li, Yanfei Kang, Jinyang Fan, et al. Macroscopic experimental study and microscopic phenomenon analysis of damage self-healing in salt rock[J]. Engineering Geology, 2024, 338: 107634.
59. Fritz Crotogino. Traditional bulk energy storage—Coal and underground natural gas and oil storage. Elsevier, 2022: 633–649.
60. Katarzyna Cyran. Insight into a shape of salt storage caverns[J]. Archives of Mining Sciences, 2020, 65(2).
61. P Renoux. Geological and Geophysical Study of an Hydrocarbon Storage in Salt Caverns in Manosque (France)[A]//European Association of Geoscientists & Engineers, 2013: 348.
62. Paula-D Weber, Gutierrez Karen-A, Lord David-L, et al. Analysis of SPR salt cavern remedial leach program 2013. Sandia National Lab.(SNL-NM), Albuquerque, NM (United States); GRAM, Inc ..., 2013.
63. Ioan Iordache, Schitea Dorin, Gheorghe Adrian-V, et al. Hydrogen underground storage in Romania, potential directions of development, stakeholders and general aspects[J]. international journal of hydrogen energy, 2014, 39(21): 11071–11081.
64. M Langer. Use of solution-mined caverns in salt for oil and gas storage and toxic waste disposal in Germany[J]. Engineering geology, 1993, 35(3-4): 183-190.
65. P Vidal-Nunes, Da Gama C-Dinis. Underground gas storage in Portuguese salt caverns[A]// ISRM, 2014: ISRM-EUROCK.
66. Mandhapati Raju, Khaitan Siddhartha-Kumar. Modeling and simulation of compressed air storage in caverns: a case study of the Huntorf plant[J]. Applied energy, 2012, 89(1): 474-481.
67. Xinjing Zhang, Li Yang, Gao Ziyu, et al. Overview of dynamic operation strategies for advanced compressed air energy storage[J]. Journal of Energy Storage, 2023, 66 107408.
68. Xilin Shi, Xinxing Wei, Chunhe Yang, et al. Problems and Countermeasures for Construction of China's Salt Cavern Type Strategic Oil Storage[J]. Bulletin of Chinese Academy of Sciences, 2023, 38(1): 99-111.
69. Chunhe Yang, Tao He, Wang Tongtao. Research and development progress of oil and gas storage construction technology in bedded salt rock formation[J]. Oil & Gas Storage and Transportation, 2022, 41(6): 614-624.
70. Chun-he Yang, LIANG Wei-guo, WEI Dong-hou, et al. Investigation on possibility of energy storage in salt rock in China[J]. Chinese Journal of Rock Mechanics and Engineering, 2005, 24(24): 4409–4417.

References

71. Guangkuo Li, Wang Guohua, XUE Xiaodai, et al. Design and Analysis of Condenser Mode for Jintan Salt Cavern Compressed Air Energy Storage Plant of China[J]. Automation of Electric Power Systems, 2021, 45(19): 91–99.
72. Chen Jie, Jiang Deyi, Liu Wei, et al. Research Progress of Solution Mining and Comprehensive Utilization of Salt Cavern[J]. Bulletin of National Natural Science Foundation of China, 2021, 35(6): 911-916.
73. Chunhe Yang, Wang Tongtao. Advance in deep underground energy storage[J]. Chinese Journal of Rock Mechanics and Engineering, 2022, 41(9): 1729-1759.
74. Guimin Zhang, Zhenshuo Wang, Yuxuan Liu, et al. Research on stability of the key roof above horizontal salt cavern for compressed air energy storage[J]. Rock and soil mechanics, 2021, 42(3): 800-812.
75. Xilin Shi, Qinglin Chen, Hongling Ma, et al. Geomechanical investigation for abandoned salt caverns used for solid waste disposal[J]. Bulletin of Engineering Geology and the Environment, 2021, 801205–1218.
76. Jie Yang, Zhengyou Liu, Chunhe Yang, et al. Mechanical and microstructural properties of alkali wastes as filling materials for abandoned salt caverns[J]. Waste and biomass valorization, 2021, 121581–1590.
77. Glenn-M Duyvestyn, Davidson Brett-C, Dusseault Maurice-B. Salt solution caverns for petroleum industry toxic granular solid waste disposal[A]//SPE, 1998: 47250.
78. Yiwei Ren, Yuan Qiang, Kang Yanfei, et al. Experimental Determination of Polycrystalline Salt Rock Thermal Conductivity, Diffusivity and Specific Heat From 20 to 240° C[J]. Frontiers in Earth Science, 2022, 10835974.
79. Hafssa Tounsi, Lerche Svetlana, Wolters Ralf, et al. Impact of the compaction behavior of crushed salt on the thermo-hydro-mechanical response of a generic salt repository for heat-generating nuclear waste[J]. Engineering Geology, 2023, 107217.
80. Hafssa Tounsi, Rutqvist Jonny, Hu Mengsu, et al. Long-term sinking of nuclear waste canisters in salt formations by low-stress creep at high temperature[J]. Acta Geotechnica, 2023, 1–16.
81. Pedro-ALP Firme, Roehl Deane, Romanel Celso. Salt caverns history and geomechanics towards future natural gas strategic storage in Brazil[J]. Journal of Natural Gas Science and Engineering, 2019, 72103006.
82. Peng Li, Li Yinping, Shi Xilin, et al. Stability analysis of U-shaped horizontal salt cavern for underground natural gas storage[J]. Journal of Energy Storage, 2021, 38102541.
83. Jinyang Fan, Wei Liu, Deyi Jiang, et al. Time interval effect in triaxial discontinuous cyclic compression tests and simulations for the residual stress in rock salt[J]. Rock Mechanics and Rock Engineering, 2020, 534061–4076.
84. Cormier V F, Bergman M I, Olson P L. Earth's core: geophysics of a planet's deepest interior[M]. Elsevier, 2021.
85. Xiao-Cheng Zhang, Gong Jian-Guo, Xuan Fu-Zhen. A deep learning based life prediction method for components under creep, fatigue and creep-fatigue conditions[J]. International Journal of Fatigue, 2021, 148106236.
86. Jinlong Li, Chunhe Yang, Xilin Shi, et al. Construction modeling and shape prediction of horizontal salt caverns for gas/oil storage in bedded salt[J]. Journal of Petroleum Science and Engineering, 2020, 190107058.
87. Zhaofei Chu, Zhijun Wu, ZhiYang Wang, et al. Micro-mechanism of brittle creep in saturated sandstone and its mechanical behavior after creep damage[J]. International Journal of Rock Mechanics and Mining Sciences, 2022, 149104994.
88. Lei Zhang, Hongwei Zhou, Xiangyu Wang, et al. A triaxial creep model for deep coal considering temperature effect based on fractional derivative[J]. Acta Geotechnica, 2022, 1–13.
89. Qiang Zhang, Zhanping Song, Junbao Wang, et al. Creep properties and constitutive model of salt rock[J]. Advances in Civil Engineering, 2021, 20211–29.
90. N Erarslan, Williams D-J. Mechanism of rock fatigue damage in terms of fracturing modes[J]. International Journal of Fatigue, 2012, 4376–89.
91. P-B Attewell, Farmer I-W. Fatigue behaviour of rock[A]//Elsevier, 1973: 1–9.

92. V-V Adushkin, Spivak A-A, Kharlamov V-A. Effects of lunar-solar tides in the variations of geophysical fields at the boundary between the Earth's crust and the atmosphere[J]. Izvestiya, Physics of the Solid Earth, 2012, 48104–116.
93. FINN Surlyk, Noe-Nygaard NANNA. Sand remobilisation and intrusion in the Upper Jurassic Hareelv Formation of East Greenland[J]. Bulletin of the Geological Society of Denmark, 2001, 48169–188.
94. Chuang Liu, Huamin Li, Mitri Hani. Effect of strata conditions on shield pressure and surface subsidence at a longwall top coal caving working face[J]. Rock Mechanics and Rock Engineering, 2019, 521523–1537.
95. Yi Liu, Feng Dai. A review of experimental and theoretical research on the deformation and failure behavior of rocks subjected to cyclic loading[J]. Journal of Rock Mechanics and Geotechnical Engineering, 2021, 13(5): 1203-1230.
96. Xiurun Ge, Yu Jiang, Yunde Lu, et al. Testing study on fatigue deformation law of rock under cyclic loading [J]. Chinese Journal of Rock Mechanics and Engineering, 2003, 22(10): 1581–1585.
97. Yintong Guo, Kelie Zhao, Guanhua Sun, et al. Experimental study of fatigue deformation and damage characteristics of salt rock under cyclic loading[J]. Rock and soil mechanics, 2011, 32(5): 1353–1359.
98. Linjian Ma, Xinyu Liu, Hongfa Xu, et al. Experimental study on triaxial deformation and strength characteristics of salt rock under cyclic loads [J]. Chinese Journal of Rock Mechanics and Engineering, 2013, 32(4): 849-856.
99. Yunfeng Zhao, Ren Song, Jiang Deyi, et al. Influence of wetting-drying cycles on the pore structure and mechanical properties of mudstone from Simian Mountain[J]. Construction and Building Materials, 2018, 191923–931.
100. 100. Yanbo Zhu, Xing Huang, Jie Guo, et al. Experimental study of fatigue characteristics of gypsum rock under cyclic loading[J]. Chinese Journal of Rock Mechanics and Engineering, 2017, 36(4): 940-952.
101. Shuhong Wang, Zihe Wang, Kaiyi Wang, et al. Evolution Law of Elastic Modulus of Sandstone with Double Fissures Under Cyclic Loading[J]. Journal of Northeastern University(Natural Science), 2020, 41(2): 282–286.
102. Ming Gao Lu, Yuanhui Li. Influence of confining pressure on fatigue deformation properties of yellow sandstone[J]. Rock and soil mechanics, 2016, 37(7): 1847–1856.
103. Yue Han, Hongling Ma, Chunhe Yang, et al. A modified creep model for cyclic characterization of rock salt considering the effects of the mean stress, half-amplitude and cycle period[J]. Rock Mechanics and Rock Engineering, 2020, 533223–3236.
104. Jinyang Fan, Jie Chen, Deyi Jiang, et al. Discontinuous cyclic loading tests of salt with acoustic emission monitoring[J]. International Journal of Fatigue, 2017, 94140–144.
105. Deyi Jiang, JinYang Fan, Jie Chen, et al. A mechanism of fatigue in salt under discontinuous cycle loading[J]. International Journal of Rock Mechanics and Mining Sciences, 2016, 86255–260.
106. Feng Chen, Chunhe Yang, Shiwei Bai. Investigation on optimized gas recovery velocity of natural gas storage in salt rock layer by numerical simulation[J]. Rock and soil mechanics, 2007, 28(1): 57–62.
107. Hongwei Zhou, Jinming He, Zhide Wu. Permeability properties of interbedded saline rocks and their fine structure characterization [J]. Chinese Journal of Rock Mechanics and Engineering, 2009, 28(10): 2068–2073.
108. Hongbo Gao, Weiguo Liang, Suguo Xu, et al. Study on the response of mechanical properties of salt rock under cyclic loads [J]. Chinese Journal of Rock Mechanics and Engineering, 2011, (Z1).
109. Feng He, Song Yang. Study on the viscoelastic-plastic fractional creep model of sandstone in uniaxial compression[J]. Chinese Journal of Applied Mechanics, 2022, 1–8.
110. Ming Huang, Xinrong Liu, Yunhua Zhu, et al. A study of behaviors of generalized Kelvin-Voigt model under low frequency cyclic load[J]. Rock and soil mechanics, 2009, 30(8): 2300–2304.

References

111. Qiangguo Jian. Study on the Criterion of Rock Salt Based on Energy Principles and its Application in Engineering[D]. Chongqing university, 2014.
112. Benjamin Cerfontaine, Charlier Robert, Collin Frédéric, et al. Validation of a new elasto-plastic constitutive model dedicated to the cyclic behaviour of brittle rock materials[J]. Rock Mechanics and Rock Engineering, 2017, 502677–2694.
113. Qiang Zhang, Junbao Wang, Zhanping Song, et al. Microstructure variation and empirical fatigue model of salt rock under cyclic loading[J]. Rock and soil mechanics, 2022, 43(4): 995–1008.
114. Yongjie Yang Yang, Song, Jun Chu. Experimental study on strength and deformation characteristics of coal rock under cyclic loading effects [J]. Chinese Journal of Rock Mechanics and Engineering, 2007, 26(1): 201–205.
115. Yi Liu, Dai Feng. A damage constitutive model for intermittent jointed rocks under cyclic uniaxial compression[J]. International Journal of Rock Mechanics and Mining Sciences, 2018, 103289–301.
116. Shu chun Li, JiangXu, Yunqi Tao, et al. Low cycle fatigue damage model and damage variable expression of rock[J]. Rock and soil mechanics, 2009, 30(6): 1611–1614, 1619.
117. C-H Scholz. Mechanism of creep in brittle rock[J]. Journal of Geophysical Research, 1968, 73(10): 3295–3302.
118. Q-Y Wang, Zhu W-C, Xu T, et al. Numerical simulation of rock creep behavior with a damage-based constitutive law[J]. International Journal of Geomechanics, 2017, 17(1): 4016044.
119. E Maranini, Brignoli M. Creep behaviour of a weak rock: experimental characterization[J]. International Journal of Rock Mechanics and Mining Sciences, 1999, 36(1): 127–138.
120. Nicolae Cristescu, Hunsche Udo. Time effects in rock mechanics[M]. Wiley New York, 1998.1
121. H Ito, Sasajima S. A ten year creep experiment on small rock specimens[A]//Elsevier, 1987: 113–121.
122. David Griggs. Creep of rocks[J]. The Journal of Geology, 1939, 47(3): 225–251.
123. Dehong Li, Mingyuan Yu, Dapeng Tian, et al. Study on Creep Law and Nonlinear Creep Model of Serpentine Marble[J]. Chinese Journal of Underground Space and Engineering, 2023, 19(2): 420–427.
124. Xunjian Hu, Kang Bian, Jian Liu, et al. Discrete element simulation study on the influence of microstructure heterogeneity on the creep characteristics of granite[J]. Chinese Journal of Rock Mechanics and Engineering, 2019, 38(10): 2069–2083.
125. Changbing Zhou, Zhijun Wan, Yuan Zhang, et al. Creep characteristics and constitutive model of gas coal mass under high temperature and triaxial stress[J]. Journal of China Coal Society, 2012, 37(12): 2020–2025.
126. Yuan Zhang, YalingWang, Jin Yu, et al. Creep behavior and its nonlinear creep model of deep gypsum mudstone[J]. Rock and soil mechanics, 2018, (s1).
127. Sana Zafar, Hedayat Ahmadreza, Moradian Omid. Micromechanics of fracture propagation during multistage stress relaxation and creep in brittle rocks[J]. Rock Mechanics and Rock Engineering, 2022, 55(12): 7611–7627.
128. C-J Spiers, Peach C-J, Brzesowsky R-H, et al. Long-term rheological and transport properties of dry and wet salt rocks. Commission of the European Communities, 1988.
129. Diansen Yang, Chen Weizhong, Yang Jianping, et al. Application of digital image correlation technique in experimental study of the creep behavior and time dependent damage of natural rock salt[J]. Journal of Testing and Evaluation, 2012, 40(2): 220–226.
130. Yue Ma, Mian Chen, Chunhe Yang. Study on the effect of gypsum content on the creep rate of salt paste layer [J]. Chinese Journal of Rock Mechanics and Engineering, 2013, (z2): 3238–3244.
131. Lance-A Roberts, Buchholz Stuart-A, Mellegard Kirby-D, et al. Cyclic loading effects on the creep and dilation of salt rock[J]. Rock Mechanics and Rock Engineering, 2015, 482581–2590.
132. Wenjing Li, Han Yanhui,Wang Tao, et al. DEM micromechanical modeling and laboratory experiment on creep behavior of salt rock[J]. Journal of Natural Gas Science and Engineering, 2017, 4638–46.
133. Amornrat Luangthip, Wilalak Naphaphat, Thongprapha Thanittha, et al. Effects of carnallite content on mechanical properties of Maha Sarakham rock salt[J]. Arabian Journal of Geosciences, 2017, 101–14.

134. M Hamami. Experimental and numerical studies of rock salt strain hardening[J]. Geotechnical & Geological Engineering, 2006, 241271–1292.
135. Jianqing He, Mengyuan Lin, Liguo Chen, et al. Empirical creep model for K0 consolidation of lacustrine soft clay considering confining pressure[J]. Journal of Natural Disasters, 2022, 31(1): 147–156.
136. Baodi Wang, Liping Su, Yang Liu. Improved Nishihara creep model research and secondary development calculation[J]. Journal of Safety Science and Technology, 2022, 18(8): 128–134.
137. Juntao Yang, Yanqi Song, Hongfa Ma, et al. A creep constitutive model of salt rock considering hardening and damage effects[J]. Rock and soil mechanics, 2023, (10): 1–14.
138. Lixin Wu, JinzhuangWang. A preliminary study of the rheological properties of coal rocks and their microscopic influence characteristics [J]. Chinese Journal of Rock Mechanics and Engineering, 1996, 4(4).
139. Jingyang Ding, Hongwei Zhou, Chao Li, et al. Fractional order creep intrinsic modeling of salt rock based on Weibull distribution [J]. Chinese Journal of Solid Mechanics, 2013, 34(5): 8.
140. Xiangdong Zhang, Qiang Fu. Study on triaxial creep test of mudstone[J]. Chinese Journal of Applied Mechanics, 2012, 29(2): 154–158.
141. You Wang, Xiaoyu Wang, Yaqian Li. Studies on the Improved Generalized Kelvin Rock Creep Model[J]. Journal of Hebei University of Engineering(Natural Science Edition), 2020, 37(4): 47–51.
142. Yang Han, Yuehu Tan, Erbing Li, et al. Nonconstant Burgers creep model for rocks and its parameter identification [J]. Engineering Mechanics, 2018, 35(3): 210–217.
143. Qianfeng Yin, Haifeng Lu, Fengchun Zhao. A Plasticity - based Improved Maxwall Model of Rock Creep[J]. Journal of Tangshan University, 2017, 30(03): 55–57.
144. Haifei Jiang, Dongyan Liu, Wei Huang, et al. Creep properties of rock under high confining pressure and different pore water pressures and a modified Nishihara model[J]. Chinese Journal of Geotechnical Engineering2014, 36(03): 443–451.
145. Qinghui Jiang, Yajing Qi, ZhijianWang, et al. An extended Nishihara model for the description of three stages of sandstone creep[J]. Geophysical Journal International, 2013, 193(2): 841–854.
146. Anrun Li, Deng Hui, Zhang Haojie, et al. Developing a two-step improved damage creep constitutive model based on soft rock saturation-loss cycle triaxial creep test[J]. Natural Hazards, 2021, 108(2): 2265–2281.
147. Ehsan Haghighat, Rassouli Fatemeh-S, Zoback Mark-D, et al. A viscoplastic model of creep in shale[J]. Geophysics, 2020, 85(3): MR155–MR166.
148. Linjian Ma,Wang MingYang, Zhang Ning, et al. A variable-parameter creep damage model incorporating the effects of loading frequency for rock salt and its application in a bedded storage cavern[J]. Rock Mechanics and Rock Engineering, 2017, 502495–2509.
149. Aditya Singh, Kumar Chandan, Kannan L-Gopi, et al. Engineering properties of rock salt and simplified closed-form deformation solution for circular opening in rock salt under the true triaxial stress state[J]. Engineering Geology, 2018, 243218–230.
150. Kui Wu, Shao Zhushan, Sharifzadeh Mostafa, et al. Analytical approach to estimating the influence of shotcrete hardening property on tunnel response[J]. Journal of Engineering Mechanics, 2022, 148(1): 4021127.
151. Rongbin Hou, Kai Zhang, Jing Tao, et al. A nonlinear creep damage coupled model for rock considering the effect of initial damage[J]. Rock Mechanics and Rock Engineering, 2019, 521275–1285.
152. Hao Cheng, YuChen Zhang, Xiaoping Zhou. Nonlinear creep model for rocks considering damage evolution based on the modified Nishihara model[J]. International Journal of Geomechanics, 2021, 21(8): 4021137.
153. HongWei Zhou, ChunPing Wang, ZhiQiang Duan, et al. Time-based fractional derivative approach to creep constitutive model of salt rock[J]. SCIENTIA SINICA Physica,Mechanica & Astronomica, 2012, 42(3): 310–318.
154. XingangWang, Yin Yueping,Wang Jiading, et al. A nonstationary parameter model for the sandstone creep tests[J]. Landslides, 2018, 151377–1389.

References

155. Yibin Zhang, Yihai Zhang, Haitao Ma, et al. Study on nonlinear viscoelasto-plastic damage creep model of rock and its parameter identification[J]. Journal of Safety Science and Technology, 2023, 19(6): 13–19.
156. Bingbing Yu, Qing Li, Tongde Zhao, et al. Full-time nonlinear creep damage model of fractured rock mass based on stress-time double threshold[J]. Chinese Journal of Rock Mechanics and Engineering, 2023, 1–17.
157. H-W Zhou, C-PWang, B-B Han, et al. A creep constitutive model for salt rock based on fractional derivatives[J]. International Journal of Rock Mechanics and Mining Sciences, 2011, 48(1): 116–121.
158. Youliang Chen, Qijian Chen, Yungui Pan, et al. A Chemical Damage Creep Model of Rock Considering the Influence of Triaxial Stress[J]. Materials, 2022, 15(21): 7590.
159. Dashnor Hoxha, Homand Françoise, Auvray Christophe. Deformation of natural gypsum rock: Mechanisms and questions[J]. Engineering Geology, 2006, 86(1): 1–17.
160. Xiaozhao Li, Qishuo Zhang, Fucong Chai, et al. Modeling of static creep fracture of brittle rock after dynamic damage [J]. Chinese Journal of Theoretical and Applied Mechanics, 2023, 55(04): 903–914.
161. Xiulei Li, Qiwei Li, Qian Li. A study of the creep model of rock considering fractures and thermal damage[J]. Hydrogeology and Engineering Geology, 2019, 46(6): 46–56.
162. Sunggi Baik, Raj R. Mechanisms of creep-fatigue interaction[J]. Metallurgical Transactions A, 1982, 131215–1221.
163. Jean Lemaitre, Desmorat Rodrigue. Engineering damage mechanics: ductile, creep, fatigue and brittle failures[M]. Springer Science & Business Media, 2006.
164. Hongyin Mao, Mahadevan Sankaran. Reliability analysis of creep–fatigue failure[J]. International journal of fatigue, 2000, 22(9): 789–797.
165. P Rodriguez, Rao K-Bhanu-Sankara. Nucleation and growth of cracks and cavities under creep-fatigue interaction[J]. Progress in materials science, 1993, 37(5): 403–480.
166. M Sauzay, Mottot M, Allais L, et al. Creep-fatigue behaviour of an AISI stainless steel at 550 C[J]. Nuclear Engineering and Design, 2004, 232(3): 219–236.
167. R-P Skelton, Gandy D. Creep–fatigue damage accumulation and interaction diagram based on metallographic interpretation of mechanisms[J]. Materials at High Temperatures, 2008, 25(1): 27–54.
168. Yasuhide Asada, Dozaki Koji, Ueta Masahiro, et al. Exploratory research on creep and fatigue properties of 9Cr-steels for the steam generator of an FBR[J]. Nuclear engineering and design, 1993, 139(3): 269–275.
169. ZhenWang, Wu WenWang, Liang Jiecun, et al. Creep–fatigue interaction behavior of nickel-based single crystal superalloy at high temperature by in-situ SEM observation[J]. International Journal of Fatigue, 2020, 141105879.
170. Sunil Goyal, Mariappan K, Shankar Vani, et al. Studies on creep-fatigue interaction behaviour of Alloy 617M[J]. Materials Science and Engineering: A, 2018, 73016–23.
171. Hongzhou Zhu, Heng Yan, Boming Tang. Damage Model of Interaction Between Fatigue and Creep for Asphalt Mixture[J]. China Journal of Highway and Transport, 2011, 24(04): 15–20.
172. Jiali Qi. Research on cumulative deformation of saturated soft clay considering the creep effect under long-term cyclic loading [D]. Tianjin University, 2016.
173. Xiaxing Zhou, Dagang Li. Fracture Mechanism of Plastic-wood Floors under Fatigue and Creep Interaction[J]. FORESTRY MACHINERY & WOODWORKING EQUIPMENT, 2009, 37(03): 21–23.
174. M-M Sain, Balatinecz J, Law S. Creep fatigue in engineered wood fiber and plastic compositions[J]. Journal of Applied Polymer Science, 2000, 77(2): 260–268.
175. Deyi Jiang, Zhenyu Yang, JinYang Fan. et al. Experimental study of load rate effect of salt rock during loading and unloading[J]. Rock and soil mechanics, 2023, 44(2): 403–414.
176. Junbao Wang, Xinrong Liu, Ming Huang. et al. Analysis of axial creep properties of salt rock under low frequency cyclic loading using Burgers model[J]. Rock and soil mechanics, 2014, (4): 933–942.

177. Sheng-jun Miao, Pengjin Yang, Hui Wang, et al. Fatigue rheological damage modeling of siltstone under cyclic loads [J]. Engineering Mechanics, 2022, 39(7): 70–80.
178. Hongfa Xu, Liangliang Qi, Bin Liu, et al. Poyhting-Thomson model of rock salt under cyclic loading[J]. Journal of vibration and shock, 2018, 37(13): 203–209.
179. Zong-ze Li, Deyi Jiang, JinYang Fan, et al. Experimental study of triaxial interval fatigue of salt rock[J]. Rock and soil mechanics, 2020, 41(4): 1305–1312, 1322.
180. Deyi Jiang, Jinyang Fan, Jie Chen, et al. Influence of interval fatigue tests on fatigue characteristics of salt rock[J]. Chinese Journal of Geotechnical Engineering, 2016, 38(7): 1181–1186.
181. Zongze Li, Jinyang Fan, Marion Fourmeau, et al. Long-term deformation of rock salt under creep–fatigue stress loading paths: Modeling and prediction[J]. International Journal of Rock Mechanics and Mining Sciences, 2024, 181: 105861.
182. Yao Cui, Deyi Jiang, Fengbin DU, et al. Experimental study on character of acoustic emission caused by interval fatigue of salt rock[J]. Journal of Central South University (Science and Technology), 2017, 48(7): 1875–1882.
183. Deyi Jiang, Wenhao Liu, Jie Chen, et al. Fatigue performance of ordinary concrete subjected to stepwise discontinuous cyclic loading[J]. Journal of Southeast University(Natural Science Edition), 2019, 49(4): 631–637.
184. Ju Wang, Jiangteng Li, Zhanming Shi. Deformation damage and acoustic emission characteristics of red sandstone under fatigue–creep interaction[J]. Theoretical and Applied Fracture Mechanics, 2022, 117103192.
185. Zhanming Shi, Jiangteng Li, JuWang. Effect of creep load on fatigue behavior and acoustic emission characteristics of sandstone containing pre-existing crack during fatigue loading[J]. Theoretical and Applied Fracture Mechanics, 2022, 119103296.
186. Linjian Ma, Yunxiao Wang, Ming Yang, Wang et al. Mechanical properties of rock salt under combined creep and fatigue[J]. International Journal of Rock Mechanics and Mining Sciences, 2021, 141104654.
187. Kai Zhao, Ma Hongling, Xiaopeng Liang, et al. Damage evaluation of rock salt under multi-level cyclic loading with constant stress intervals using AE monitoring and CT scanning[J]. Journal of Petroleum Science and Engineering, 2022, 208109517.

Open Access This chapter is licensed under the terms of the Creative Commons Attribution-NonCommercial-NoDerivatives 4.0 International License (http://creativecommons.org/licenses/by-nc-nd/4.0/), which permits any noncommercial use, sharing, distribution and reproduction in any medium or format, as long as you give appropriate credit to the original author(s) and the source, provide a link to the Creative Commons license and indicate if you modified the licensed material. You do not have permission under this license to share adapted material derived from this chapter or parts of it.

The images or other third party material in this chapter are included in the chapter's Creative Commons license, unless indicated otherwise in a credit line to the material. If material is not included in the chapter's Creative Commons license and your intended use is not permitted by statutory regulation or exceeds the permitted use, you will need to obtain permission directly from the copyright holder.

Chapter 2
Dilatancy Properties of Salt Under Monotonous Compression and Brief Introduction into Dislocation Theory

Underground salt caverns form during exploiting NaCl sedimentary. NaCl crystal is isometric system hexoctahedron symmetry class. Monocrystal appears hexahedron, typical structure of AX type chemical compound, where negative ions are located in angular point and face center of cubic hail and close piling up and positive ions fill the hole inside the hexahedrons. Fresh surface of NaCl crystal appears metallic luster. NaCl crystal easily deliquesces and has full cube cleavage plane (Fig. 2.1).

Salt deposit is one kind of chemical sedimentary deposit forming by vaporizing and eliminating water in dry climate conditions during geological process. Deposition type of ore deposit occurs as bedded and contains impurities of mudstone in some situations. Because of the tectonic movement, rock salt layer flows upward, intrudes into overlaying rock or hog overlaying rock, forming salt domes. Most of rock salt layer have simple morphology and are bedded.

Salt cavern construction as the ideal reservation site of oil and natural gas and subterranean disposal site of highly active nuclear waste is exploited and utilized with various risk accompanied, such as, volume shrinkage of cavity, surface subsidence et al [1, 2]. In this chapter, investigation into the basic properties is performed to lay the foundation for the subsequent research presented in the following chapters.

2.1 Experimental Materials and Methods

2.1.1 Rock Salt Material and Specimens

Rock salt is a sedimentary rock composed mainly of NaCl, with metallic glossy on fresh surfaces, easily absorbed by water via deliquescence, and with a cubic crystal form. Salt deposits are chemically deposited by the evaporation and concentration of water under arid climatic conditions and specific geological conditions. Usually,

Fig. 2.1 NaCl crystalline

sedimentary deposits are stratified and in some cases contain impurities and inclusions, such as mudstone. Due to tectonic movement, a buried rock salt layer will flow upward and intrude into the overlying rock layer or make the overlying rock layer arch up, forming salt mounds. Rock salts are widely distributed in the Earth's crust, and the largest salt mines worldwide are located mainly in North America and Asia, where they are hundreds to thousands of meters thick [3]. The rock salt samples used in this paper are extracted from the Khewra rock salt mine located in Pakistan in the northern Himalayas, excavated within the base of a thick layer of highly folded, faulted, and stretched Ediacaran to early Cambrian evaporites of the Salt Range Formation. Figure 2.2 shows the geological distribution map of rock salts in northern Pakistan [4].

This salt is translucent, pink, reddish to beef-colored red, while some exhibits alternating bands of red and white. The composition of the rock salt was observed using X-ray diffraction (XRD), and it was found to have a high NaCl content of more than 96% (Fig. 2.3). The remaining components are small amounts of insoluble impurities, such as Na_2SO_4, K_2SO_4 and mudstone.

The rock salts selected for processing were derived from large rocks buried at the same horizon and without significant differences or fractures. According to the "Rock Mechanics Test Procedure" developed by the Laboratory and Field Test Standardization Committee of the International Society for Rock Mechanics (ISRM) [5], samples were taken in the same direction and processed into standard cylindrical

2.1 Experimental Materials and Methods

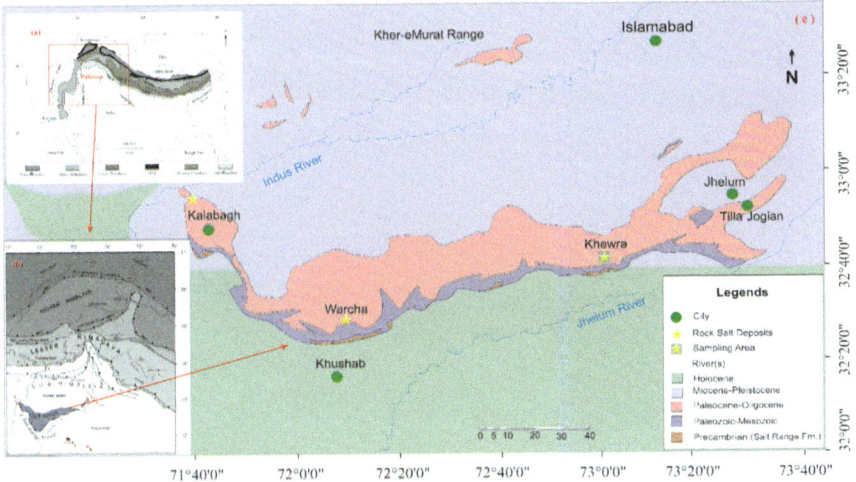

Fig. 2.2 Geological distribution map of rock salts in northern Pakistan (from [4], licensed under CC-BY 4.0)

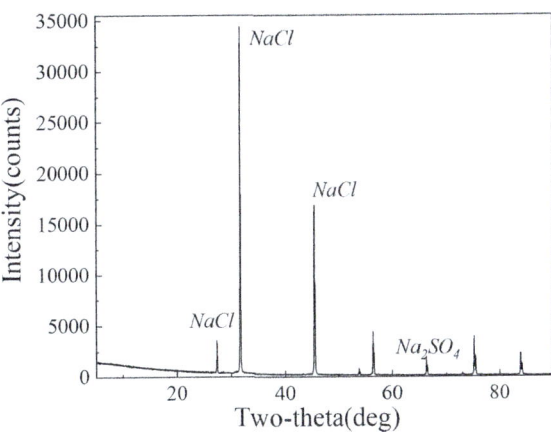

Fig. 2.3 XRD results of the tested rock salt

specimens 100 mm in height and 50 mm in diameter, with flatness errors of the upper and lower end surfaces of less than ± 0.05 mm by using a dry corer. Specimens with uniform impurity distributions, similar colors and no directly visible cracks were selected for the test. The rock salt specimens are shown in Fig. 2.4.

To mitigate the impact of moisture on the test results, the rock salt specimens were subjected to a 24-h low-temperature (60 °C) drying process in a drying oven (Fig. 2.5) prior to mechanical testing. To reduce the end effects during mechanical testing, a lubricant was applied between two Teflon shims positioned at both specimen extremities in contact with the indenter.

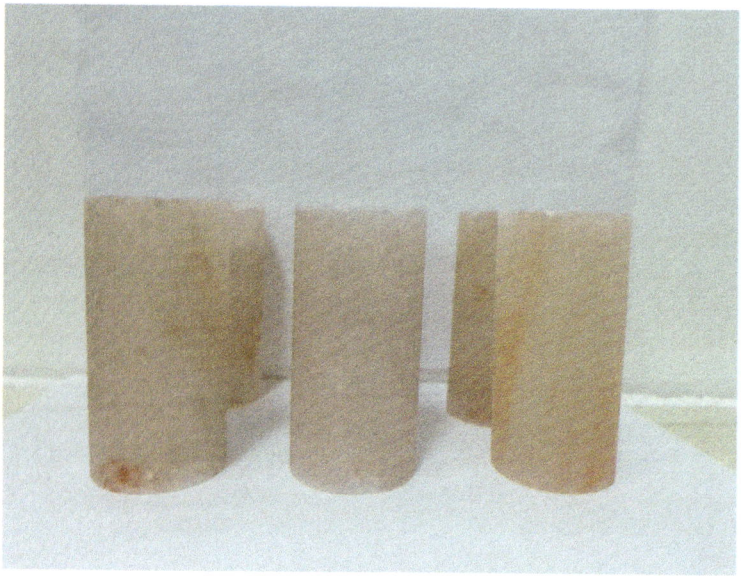

Fig. 2.4 Rock salt specimens

Fig. 2.5 Constant temperature drying oven

2.1.2 Testing Equipment

The tests were conducted at the State Key Laboratory of Coal Mine Disaster Dynamics and Control, Chongqing University, using the MTS815 Rock Mechanics Test System (Fig. 2.6), manufactured by MTS, USA.

The main technical specifications of the testing aparatus are given here:

(1) Axial Load $\leq$ 4600 kN
(2) Confining Pressure $\leq$ 140 MPa
(3) Pore Water Pressure $\leq$ 70 MPa
(4) Permeability Delta P $\leq$ 2 MPa
(5) Stiffness of Load Frame 10.5×10 N/m
(6) Hydraulic source flow 31.8 L/min
(7) Servo Valve sensitivity 290 Hz
(8) Channels of Data Acquisition 10 Chans
(9) Minimum Sampling time 50 μs
(10) Output waveforms: straight line, sine, half-sine, triangle, square, random
11) Specimens

Fig. 2.6 The MTS 815 rock mechanics test system used in this experiment

Maximum diameter of triaxial test 100 mm
Maximum height 200 mm
Maximum diameter of uniaxial test 300 mm
Maximum height 600 mm

The testing machine can realize automatic data acquisition, and is equipped with three sets of independent servo system controlling the shaft pressure, confining pressure and pore (permeability) pressure, servo valve response agile (290 HZ), high test precision; and direct contact with the specimen of the elongation instrument is also by the U.S. MTS company, can be in the high temperature (200 °C), high-pressure (140 MPa) oil in the work of the precise, can be the most accurate measurements of rock before and after the destruction of the stress–strain.

The tester can carry out uniaxial compression tests as well as triaxial tests on a variety of rocks. Stresses are defined positive in compression here. Triaxial compression tests cabe realized under various level of confining pressure, giving two possible testing conditions:

$$\left. \begin{array}{c} \sigma_1 > \sigma_2 = \sigma_3 \\ \sigma_1 = \sigma_2 > \sigma_3 \end{array} \right\} \quad (2.1)$$

Monotonic loading, fatigue and creep tests can be performed in a load- or displacement-controlled manner.

2.1.3 Test Procedure

(1) Open the test machine and place the selected rock salt samples in the center of the indenter. Apply preload to compress the specimen.
(2) Close the protective cover of the tester and turn on the tester. Set the computer program and start the test.
(3) When the samples is broken, the machine stops automatically; the power of the test system is turned off, and the specimen is removed and stored in a sealed plastic bag.

2.2 Dilatancy Features in Uniaxial Tests

2.2.1 Loading Features

To keep in consistency with the subsequent experiments, uniaxial tests employed the stress control mode, with 0.2 KN/s loading velocity. The loading continues until failure. Figure 2.7 shows the load and displacement evolution under stress control mode: after the peak, sample loses the bearing capacity and stress drops rapidly.

2.2 Dilatancy Features in Uniaxial Tests

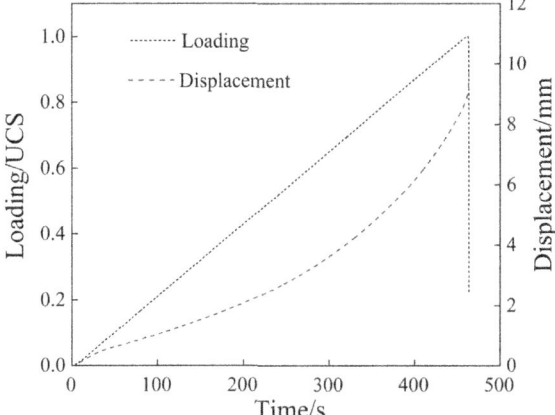

Fig. 2.7 Load, displacement path in uniaxial test

This also is the disadvantage of stress control mode: the rapid drop cannot give much time to obtain dense enough data. Since the upper and lower stress limits are fixed in fatigue tests, the stress control mode has to be applied.

The test results show that: the stress increases linearly, following the control of servo system. However, the displacement increases linearly with time only at early stages and the increase become exponential. The demarcation point between linear segment and exponential segment is ambiguous, indicating that elastic phase and plastic phase are mutually intricate.

2.2.2 Volume Expansion Features

Volumetric strain is calculated as

$$\varepsilon_v = \varepsilon_1 + 2\varepsilon_3 \tag{2.2}$$

Figure 2.8 shows the curve of axial stress-axial strain, lateral strain and volumetric strain. To avoid the superposition of radial deformation curve and volumetric deformation, the expansion of lateral strain is defined as positive (just applied in figures).

During the continuous loading the axial and radial strains increase all the time, while the volumetric strain decreases first and then increases. Three curves show the main features as follows:

Axial strain at the initial loading stage (OA segment), grows slowly, then experiences a short linear increase (AC segment). Usually the two stages were considered as compaction stage and elastic stage, respectively. Afterwards strain rate rises until failure as the stress increases.

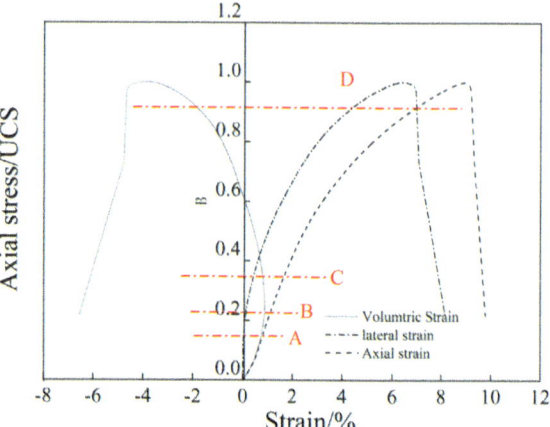

Fig. 2.8 Axial strain, lateral strain and volumetric strain curves

Radial strain at the initial loading stage (OA segment), is hardly detectable, until the stress goes up to a certain value (~ 8 MPa corresponding to A point). Thereafter, strain rate rapidly rises.

Volumetric strain goes down at the beginning, indicating that the sample as a whole is compacted. The reduction trend is transformed into an increase at ~ 11.5 MPa (point B in Fig. 2.9), where the demarcation point of compaction-dilatancy is located, indicating the volumetric reduction transformed into expansion. In dilatancy stage, volumetric strain increases.

In uniaxial test, there is no distinct elastic phase and elastoplastic phase. From the above description, the deformation stages can be summarized as follows.

Compaction stage. At the initial loading, sample experience a short compaction period from point O to point A, in which the axial strain develops with a slowing rate and the lateral strain is very small. Therefore, the volume reduces.

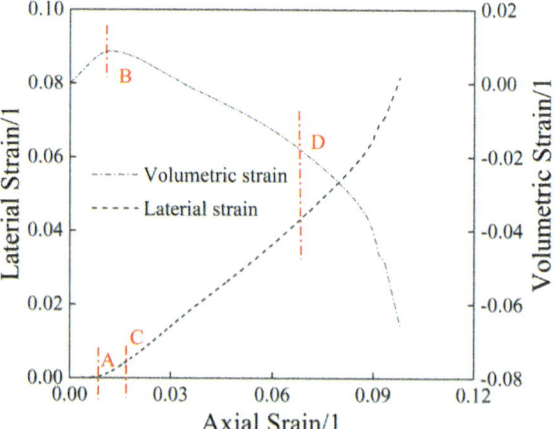

Fig. 2.9 Lateral strain and volumetric strain as a function of axial strain

2.2 Dilatancy Features in Uniaxial Tests

Composite stage. Following the compaction stage, the lateral strain increases. The sample experiences a combined effect of compaction and dilatancy, from A to C. Sample is completely compacted at point C. At this stage, the lateral strain accelerates with axial strain and stress. Since the elastic behavior of salt is linear, this stage is judged elastoplasticity which would cause dilatancy. It is the interaction of compaction and dilatancy that make the volumetric strain rate to be zero at point B, where the dilatancy point (demarcation point of compression and dilatancy) is.

Plastic stage. After complete compaction, the plastic deformation is totally due to dislocation mechanism, from point C to D. Because microcracks at this stage just finish nucleation and the scale is negligible, the lateral strain and volumetric strain develop linearly with axial strain.

Fracture stage. As the plastic strain accumulates, the connectivity of the microcracks develops which in turn influences the volumetric deformation. Cracks could produce more space inside the samples than dislocations; therefore the volumetric deformation develops faster and faster with axial strain. Once the thoroughgoing crack is formed, the failure occurs (from point D to failure).

One thing to note is that all through the deformation process is elastoplastic behavior, the elasticity and plasticity are intertwined, not mutually independent. Additionally, the segments points are distinguished by ocular estimates, in terms of the curve flatness, having strong subjectivity. The partition for every stage is not accurate, but the purpose is to understand the deformation mechanism in every stages.

2.2.3 Elastic Constants

Since the elasticity reflects the recovery properties, elastic constants are determined from unloading test. Jiang et al. [6] found that the elastic modulus and Poisson's ratio are not constant but vary with damage. As the loading proceeds in uniaxial compression test, elastic modulus increases from 6.3 to 6.7 GPa with fluctuations of ± 0.05 GPa and Poisson's ratio increases from 0.04 to 0.1. The reason for these changes is the degradation of inner particle structure. To facilitate the analysis, we assume the average values of the elastic parameters (2.3)

$$\begin{cases} E = 6.5 \text{ GPa} \\ \mu = 0.07 \end{cases} \quad (2.3)$$

E and μ are the Young's modulus and the Poisson's ratio, respectively. The shear modulus and bulk modulus can be obtained from

$$\begin{cases} G = \frac{E}{2(1+\mu)} = 3.04 \text{ GPa} \\ K = \frac{E}{3(1-2\mu)} = 2.52 \text{ GPa} \end{cases} \quad (2.4)$$

Fig. 2.10 Elasticity and plasticity for uniaxial test

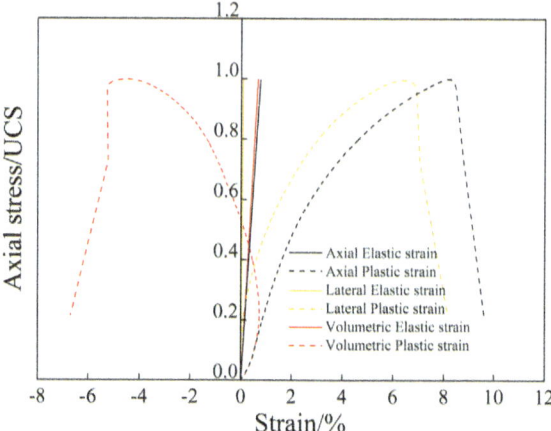

Plastic strains and elastic strain were calculated and are shown in Fig. 2.10. It is seen that the plastic curves are almost the same as in Fig. 2.8. Elastic strain is very small. This demonstrates that the plastic deformation has the overwhelming superiority among the deformation in rock salt. Therefore the description of the plasticity is fundamental for predicting the mechanical response of salt.

Dilatancy is the main mechanism causing in the volume increase. Here the generalized shear strain and shear stress are defined to obtain the parameters reflecting the dilatancy.

$$\begin{cases} \overline{\gamma} = 2\sqrt{J_2'} = \sqrt{2e_{ij}e_{ij}} = \sqrt{2\left(\varepsilon_{ij} - \frac{1}{3}\varepsilon_v\delta_{ij}\right)\left(\varepsilon_{ij} - \frac{1}{3}\varepsilon_v\delta_{ij}\right)} \\ \overline{\tau} = \sqrt{J_2} = \sqrt{s_{ij}s_{ij}} = \sqrt{\left(\sigma_{ij} - \sigma_m\delta_{ij}\right)\left(\sigma_{ij} - \sigma_m\delta_{ij}\right)} \end{cases} \quad (2.5)$$

e_{ij} and s_{ij} are the deviatoric strain and deviatoric stress tensors. ε_v and σ_m are volumetric strain and mean stress. δ_{ij} is Kronecker tensor. In uniaxial tests, generalized plastic shear strain and generalized shear stress are shown in Fig. 2.11.

Dilatancy angle is defined as show in Fig. 2.12,

$$\tan \psi = -\frac{\Delta \varepsilon_v^p}{\Delta \overline{\gamma}^p} = \beta \quad (2.6)$$

β is the dilatancy factor. Dilatancy angle initially is negative, indicating that the volume reduction then it becomes positive at point B and reaches a constant level around point C, after point D rises gradually with plastic shear strain.

At the beginning of the loading (OA segment), nothing happens with the sample. In the AC segment, the some changes at the sample surface occur, normally light red color is changed white. This is caused by the grain reduction. The new-forming crystal

2.3 Dilatancy in Triaxial Compression Test

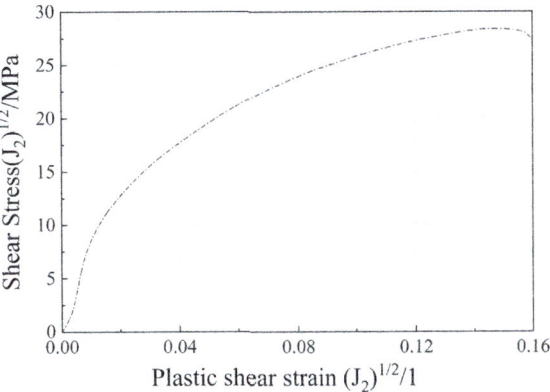

Fig. 2.11 Shear stress-plastic versus shear strain

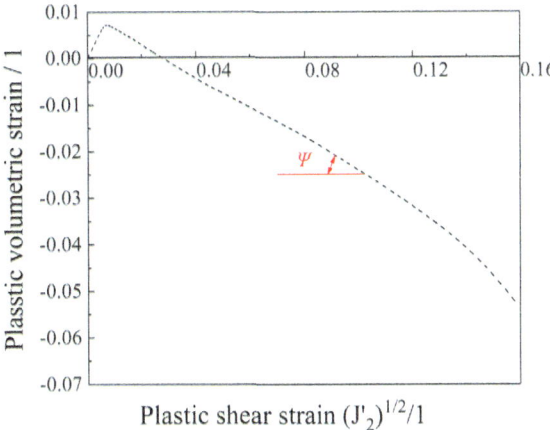

Fig. 2.12 Plastic volumetric strain versus plastic shear strain

boundaries and subboundaries block the light path for transmission. In CD segment, the sample surface does not change much. After point D, macrocrack appears on the surface and continuously grows up to cutting through the whole sample (Fig. 2.13).

2.3 Dilatancy in Triaxial Compression Test

Both in the salt cavern construction and underground excavation, the surrounding rock is in the 3-D stress state. Investigation of the properties of rocks under triaxial loading is more meaningful for engineering practice.

The same samples and loading equipment are used in triaxial compression tests. The processed samples are grouped to conduct the tests with different confining pressure. First the confining pressure is increased to the designed value with the velocity of 0.05 MPa/s, and then the axial stress is increased with the velocity of

Fig. 2.13 Specimen failed in the uniaxial test

0.2 KN/s. The tested confining pressure values are: 3, 5, 7 MPa. Because the range of the device measuring the lateral elongation is limited, the deformation cannot be measured all through the test.

The stress–strain curves from triaxial compression tests are similar to those from uniaxial compression tests, showing the same (as previously) features shown in Figs. 2.14, 2.15, 2.16, 2.17, 2.18 and 2.19. This indicates the same mechanism behind the deformation, which would be explained in the dislocation section.

Fig. 2.14 Axial strain, lateral strain and volumetric strain curves from a triaxial test with 3 MPa of confining pressure

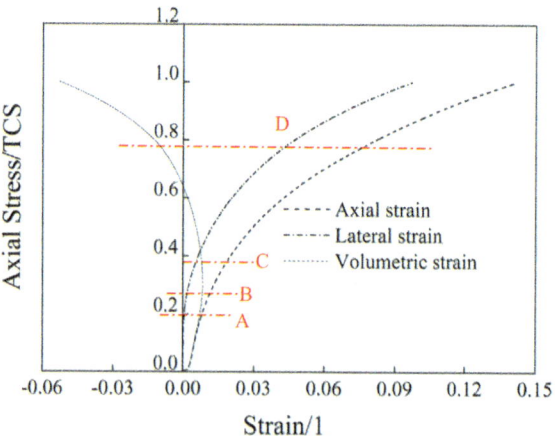

2.3 Dilatancy in Triaxial Compression Test

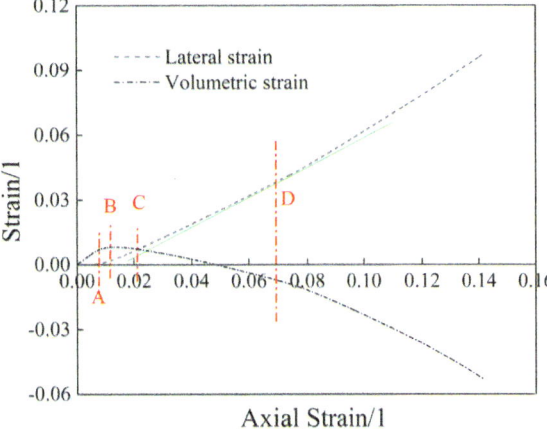

Fig. 2.15 Lateral strain and volumetric strain as a function of axial strain from a triaxial test with 3 MPa of confining pressure

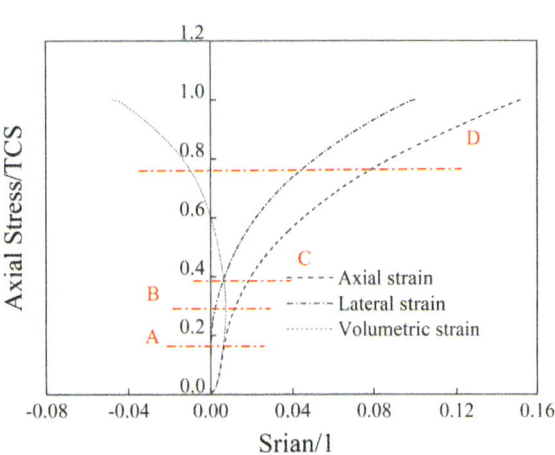

Fig. 2.16 Axial strain, lateral strain and volumetric strain curves from a triaxial test with 5 MPa of confining pressure

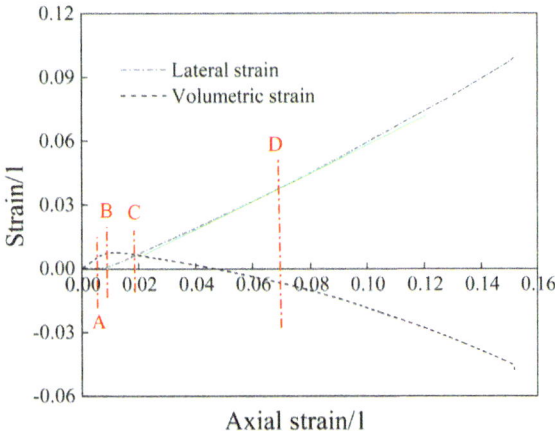

Fig. 2.17 Lateral strain and volumetric strain as a function of axial strain from a triaxial test with 5 MPa of confining pressure

Fig. 2.18 Axial strain, lateral strain and volumetric strain curves from a triaxial test with 7 MPa of confining pressure

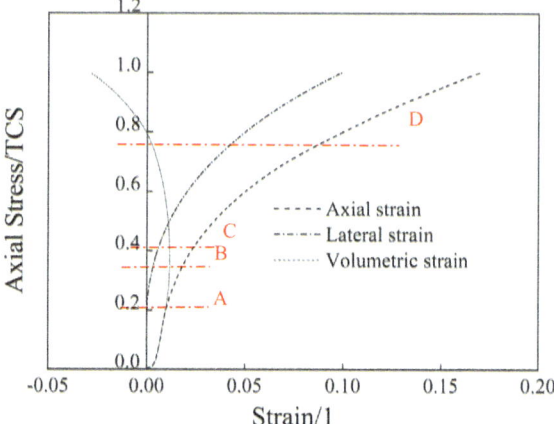

Fig. 2.19 Lateral strain and volumetric strain as a function of axial strain from a triaxial test with 7 MPa of confining pressure

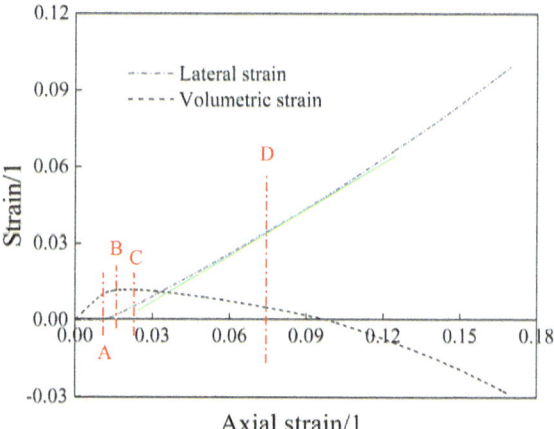

The stress–strain curves have similar characteristics of stages. The axial strains corresponding to the indicated points (see Table 2.1) are similar. Point A represents initial point of lateral strain, at around 0.75% of ε_1. Point C represents the complete compaction, at around 2% of ε_1. Point B represents the demarcation point of compression-dilatancy, where the strain rate is zero. Point D represents that the cracks run through, a new expansion mechanism, occurs 6.8 ~ 7.5% of ε_1.

With the increasing confining pressure, the constants change: both the elastic modulus and Poisson's ratio increase; the compression strength has a remarkable rise. The axial strain corresponding to the peak stress also increases significantly (see Table 2.2). All the data are obtained at room temperature 26–30°C and with the loading velocity of 0.2 KN/s.

Here, we defined the (shear-)dilatancy angle and axial-dilatancy. The tangents of them two are defined as the ratio of volumetric strain to shear strain, the ratio of volumetric strain to axial strain, respectively. Confining pressure also impacts

2.4 Dislocation Theory

Table 2.1 Axial strain corresponding to the turning points in different tests

Confining pressure σ_3/MPa	0	3	5	7
Corresponding axial strain for A (%)	0.79	0.72	0.77	1.03
Corresponding axial strain for B (%)	1.17	1.25	1.02	1.61
Corresponding axial strain for C (%)	1.76	2.01	1.86	2.10
Corresponding axial strain for D (%)	6.95	6.98	6.78	7.50

Table 2.2 Variation of elastic parameter with confining pressure

Confining pressure σ_c/MPa	0	3	5	7
Elastic modulus E/GPa	6.50	7.40	8.10	8.65
Possion's ratio μ	0.07	0.11	0.14	0.16
Compression strength σ_c/MPa	47.1	75.2	89.3	106
Axial strain corresponding to σ_c (%)	8.92	19.93	25.3	29.1

Table 2.3 Variation of shear-dilatancy angle and axial-dilatancy angle with confining pressure

Confining pressure σ_3/MPa	0	3	5	7
Shear-dilatancy angle $\tan \psi$	0.33	0.25	0.21	0.15
Axial-dilatancy angle $\tan \theta$	0.36	0.27	0.24	0.18

the growth of plastic deformation: a larger confining pressure results in a smaller (shear-)dilatancy angle ψ, axial-dilatancy θ angle and greater plastic deformation (Table 2.3).

During AD segment, under the effect of compaction and dilatancy, the salt sample mainly deformed, without visible macrocracks on the surface (Fig. 2.20). After point D, the macrocracks initiate, propagate and connect, forming the cracks (Fig. 2.21).

2.4 Dislocation Theory

2.4.1 Conceptual Framework of Dislocation

Rock salt is one kind of crystal structure, whose mechanical behavior is affected by the dislocation. As there is rarely application of dislocation theory in rock mechanics field, we briefly introduce the dislocation theory. In practice, various defects including point defect (impurity atom), planar defect, volume defect, exist in crystal and strongly change the properties of salt. Dislocation is actually a kind of line defect. There are two types of dislocations, edge dislocation and screw dislocation. Mixed dislocation is the combination of the two.

Fig. 2.20 Deformed rock salt specimen before D point from trial tests

Fig. 2.21 Fractured rock salt specimen after D point from trial tests

Dislocation theory is proposed to explain the plastic deformation. Figure 2.22 shows the deformation process of an intact crystal lattice overcoming slip resistance under the drive of shear stress.

Edge dislocation is an extra half atomic planes during gliding process, typical of dislocation, as shown in Fig. 2.23. Two categories of edge dislocation are positive and negative [7].

2.4 Dislocation Theory

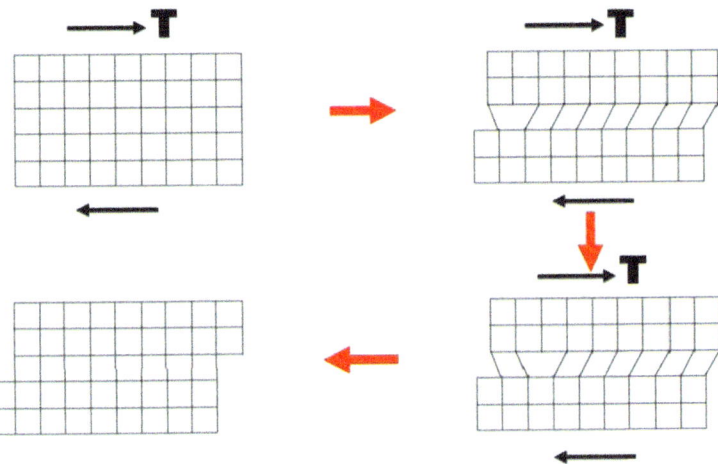

Fig. 2.22 Sliding in the crystal lattice under shear stress

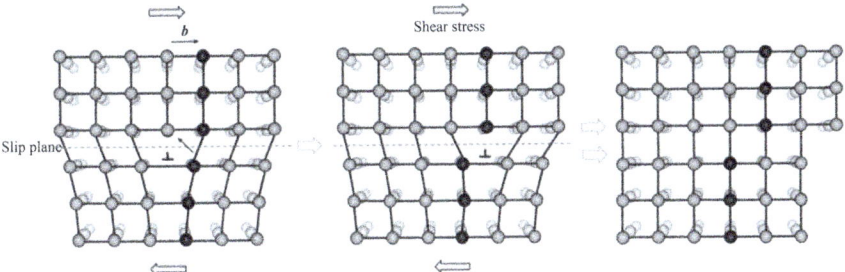

Fig. 2.23 Edge dislocation

With an applied shear stress, a side face of simple cubic crystal generates a relative scrape edge of crystal under and below the glide plane, making unaligned phenomenon around the boundary between slip zone and non-slip zone, forming a screw line surrounding the boundary. The original lattice plane cross by the boundary turns into screw surface. This lattice imperfection is screw dislocation, as shown in Fig. 2.24. Two categories of screw dislocation are left and right.

As the dislocations exists in the crystal, the atoms around the dislocation line deviate from the normal position, causing the lattice distortion and generating stress field. The stress state of that area removed the center region could be solved through the elastic theory. Dislocations cause the lattice distortion, leading to energy rise. The energy increment belongs to the dislocation strain energy, including elastic energy and core energy at the dislocation core.

Some vacancy condensation could occur during crystallization, forming dislocation sources. Three mechanisms for dislocation formation are formed by homogeneous nucleation, grain boundary initiation, and interfaces the lattice and the surface,

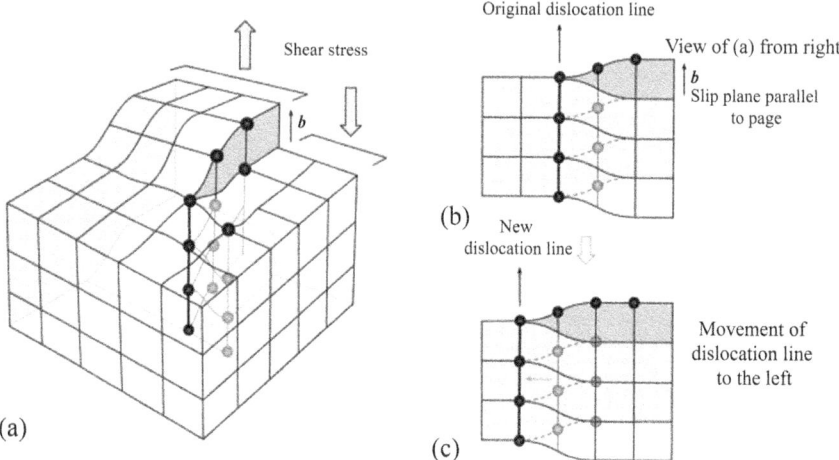

Fig. 2.24 Screw dislocation

precipitates, dispersed phases, or reinforcing fibers. Under the stress, dislocation source in the stress concentration area would ceaselessly emit dislocations to the glide direction. Once encounting obstacles, such as impurity atom, crystal boundary and subboundary, dislocations are impeded and stop gliding. As the following dislocations come, dislocation would pile up and the stress field would add together. To continue the dislocation glide, larger stress needs to apply to surmount the added stress field of pile-up dislocations. Once the number of pile-up dislocations exceeds a certain value, the crack would initiate.

2.4.2 Dislocation Behavior of Salt Under Monotonous Compression

Dislocations always form by slippage and moving of the lattice structure under shear stress. The dislocation slippage would normally leave a (set of) slip line(s). We used the same sample with the previous tests to observe the slip lines in salt. After subjected to shear stress, the sample surface was polished and then soaked in the glacial acetic acid 15 min. the observed slip lines are shown in Fig. 2.25.

The process of loading on the samples actually is an interactive process between dislocations and external force. There are numerous pores and microcracks inside the natural rock salt. In the compaction stage, salt is subjected to the external force, the original cracks are compacted while the lateral strain does not occur, this corresponds to the nonlinear elasticity, since the lateral stress is fixed. In this phase, stress concentration usually occurs around the crack tips, where the dislocation sources

2.4 Dislocation Theory 51

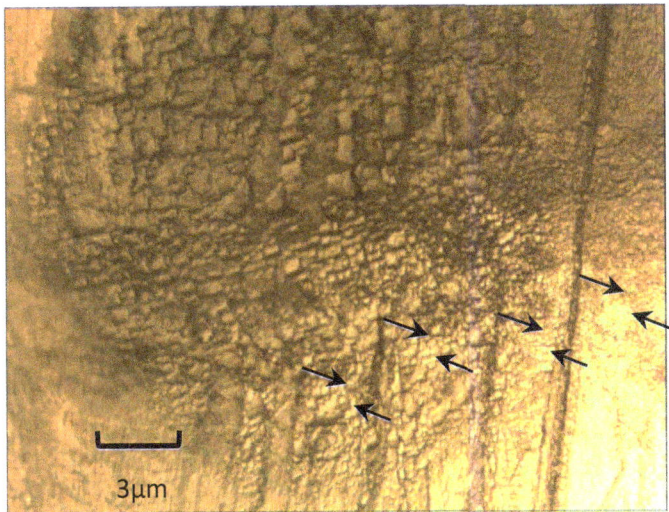

Fig. 2.25 Dislocation slide lines in the salt materials

are activated to emit dislocation, leading to irreversible deformation. Therefore the compaction is one kind of plastic behavior in some ways. This is the dislocation activities in OA segment.

In AC segment, as the stress increases, crystal structures gradually deform. Some of them are transformed into small size and a large amount of new subboundaries and boundaries form, leading to that the light-admitting quality reduces. The sample surface looks light red turning into white.

In CD segment, owing to the new crystal boundaries, the dislocations are impeded and piled up, resulting in the internal stress superposition. The impurity atoms also can block the slippage of dislocations. As the loading continues, these impediments increase external forces. Dislocation is one kind of lattice distortion leading to the volume increase. Because cracks are just in the stage of nucleation or propagate rather slowly, the dilatancy almost completely comes from dislocations, so the increment of volumetric strain is stable and linearly changes with the axial strain.

After point D, the number of plied-up dislocations comes to a critical limit. Cracks start to propagate and connect generating volume increase. In this stage, the volumetric strain increases rapidly in a geometric progression. As the lattice structures are distorted to a limit, some impurity atoms would separate out from the lattice, forming second phase. The second phase would further impede the dislocation and increase the internal stress. The cracks reduce the effective bearing area. When the average stress applied on the effective bearing area is large enough to destroy all the rest bearing element, the peak stress arrives. After peak, the cracks continue propagating and effective bear element continue reducing, resulting in a larger stress concentration.

As the salt samples are confined, crystal boundaries are protected better, needing more energy to open. More dislocations would traverse the crystal boundary. Therefore the plastic deformation is larger under larger confining pressure.

2.5 Conclusions

In this chapter, the dislocation theory is used to explain the behavior of salt under continuous compression. In the uniaxial compression tests the deformation is mostly plastic, the elastic deformation being very small. The dilatancy behavior of salt could be understood considering three mechanisms and four stages.

(1) Three mechanisms: Compaction: at lower stress, the pores are compacted. Dislocation: plastic deformation is induced by the dislocations causing the lattice structure distortion. Crack: at higher stress, the cracks propagate and result in the increase of internal space.
(2) Four stages: Compaction stage: dislocations move around the tips of original pores/defects and the total volume reduces. Composite stage: numerous dislocations at the stress concentration area slip and the compaction simultaneously continues. The volume development transforms from compaction to expansion. Plastic stage: the microcracks are in the nucleation, but negligible to the volumetric increment. Dislocation slipping results in plastic volumetric increment, which is linear with axial strain. Rupture stage: lattice structures distort to a critical limit. Crystal boundaries transform into cracks and greatly volume increase.
(3) The process of loading on the salt actually is an interactive process between dislocations and external force. In triaxial compression tests, as the confining pressure increase, Elastic modulus and Poisson's rate as well as the strength increase with the confining pressure σ_c. Dilatancy difficulty increases and dilatancy angle decrease. Crystal boundaries are fortified and more dislocations could traverse the boundaries, leading to a larger plastic deformation at higher σ_c.

References

1. Marion Fourmeau, Wen Liu, Zongze Li, et al. Research status of creep–fatigue characteristics of salt rocks and stability of compressed air storage in salt caverns[J]. Earth Energy Science, 2024.
2. Jinyang Fan, Fan Yang, Zongze Li, et al. Creep constitutive model for salt rock considering loading and unloading path based on state variables[J]. Geomechanics and Geophysics for Geo-Energy and Geo-Resources, 2025, 11(1): 30.
3. Oliver SChenk, Urai Janos-L. Microstructural evolution and grain boundary structure during static recrystallization in synthetic polycrystals of sodium chloride containing saturated brine[J]. Contributions to Mineralogy and Petrology, 2004, 146671–682.

References

4. Syed-Asim Hussain, Han Feng-Qing, Ma Zhe, et al. Unraveling sources and climate conditions prevailing during the deposition of neoproterozoic evaporites using coupled chemistry and boron isotope compositions ($\delta 11B$): the example of the salt range, Punjab, Pakistan[J]. Minerals, 2021, 11(2): 161.
5. C-E Fairhurst, Hudson John-A. Draft ISRM suggested method for the complete stress-strain curve for intact rock in uniaxial compression[J]. International journal of rock mechanics and mining sciences (1997), 1999, 36(3): 279–289.
6. Deyi Jiang, Fan Jin Yang, Jie Chen, et al. A mechanism of fatigue in salt under discontinuous cycle loading[J]. International Journal of Rock Mechanics and Mining Sciences, 2016, 86255–260.
7. Cormier V F, Bergman M I, Olson P L. Earth's core: geophysics of a planet's deepest interior[M]. Elsevier, 2021.

Open Access This chapter is licensed under the terms of the Creative Commons Attribution-NonCommercial-NoDerivatives 4.0 International License (http://creativecommons.org/licenses/by-nc-nd/4.0/), which permits any noncommercial use, sharing, distribution and reproduction in any medium or format, as long as you give appropriate credit to the original author(s) and the source, provide a link to the Creative Commons license and indicate if you modified the licensed material. You do not have permission under this license to share adapted material derived from this chapter or parts of it.

The images or other third party material in this chapter are included in the chapter's Creative Commons license, unless indicated otherwise in a credit line to the material. If material is not included in the chapter's Creative Commons license and your intended use is not permitted by statutory regulation or exceeds the permitted use, you will need to obtain permission directly from the copyright holder.

Chapter 3
Conventional Creep and Fatigue Mechanical Properties of Rock Salt

Compressed air energy storage (CAES) plants require periodic air injection and extraction based on energy demand [1]. During the injection process, as compressed air is continuously injected, the mechanical energy of the compressed air is stored in the salt cavern [2]. The storage capacity is largely determined by the size of the salt cavern and the maximum allowable pressure. During energy demand periods, the stored mechanical energy is converted back to electrical energy [3]. As the compressed air decreases, the internal air pressure of the salt cavern will gradually decrease to the minimum operating pressure [4]. Therefore, during the operation of the power plant, due to different operating schemes, the surrounding rock of the storage cavern will be subjected to cyclic loading, resulting in creep deformation and fatigue effects [5]. Consequently, a comprehensive understanding and investigation of the creep and fatigue characteristics of rock salt are crucial for ensuring the safe operation of CAES salt cavern storage facilities.

3.1 Experimental Methods

In the study of the creep and fatigue basic mechanical properties of rock salt, many researchers have already conducted a large number of tests. Considering the need for compressed air energy storage power stations to adjust peaks and fill valleys, the gas pressure limit and injection rate will vary with different operating schemes. Therefore, this chapter mainly investigates the creep mechanical properties of rock salt under different stress levels and the fatigue mechanical properties under different stress rates.

The creep mechanics test plan under different stress levels is as follows: The test is divided into two groups. One group is for incremental creep test (ICT), with the loading path as shown in Fig. 3.1a. The test plan is as follows: with the loading rate set to 1 kN/s, the loading stress is increased to 4 MPa and maintained constant for

2 h. Then, following the same loading rate, maintaining the same creep time, the stress is increased by 4 MPa for each level, conducting an incremental stress creep test until the specimen fails. The incremental stress test results show that the rock salt specimen failed after being maintained at a stress level of 24 MPa for 40 min. Based on these test results, in the decremental creep test (DCT), using the same loading rate, the stress level is first loaded to 20 MPa and maintained for 2 h. Then, using the same unloading rate of 1 kN/s, it is gradually reduced to 0 MPa, with each stress level maintained at 4 MPa. The loading path is shown in Fig. 3.1b.

The fatigue mechanics test plan under different stress rates is as follows: Two groups of mixed-rate cyclic loading and unloading tests are conducted. One group is for different rate mixed fatigue tests (DMT), setting 85% of the rock salt's uniaxial compressive strength, i.e., 25.5 MPa, as the upper stress limit for the stress rate test, and 3% of the uniaxial compressive strength, i.e., 1 MPa, as the lower stress limit. The test loading and unloading path is shown in Fig. 3.1c, with loading and unloading rates of 0.04 kN/s for cycle A, 0.2 kN/s for cycle B, 1 kN/s for cycle C, and 5 kN/s for cycle D. Every four cycles form a complete cycle, which is then repeated periodically. The other group is for graded mixed rate fatigue tests (GMT), with the initial upper stress limit set at 40% of the compressive strength, i.e., 12 MPa, and the lower stress limit set at 1 MPa. The test cycles follow the rules of the different rate mixed fatigue tests, repeating 5 large cycles, i.e., after 20 cycles of loading and

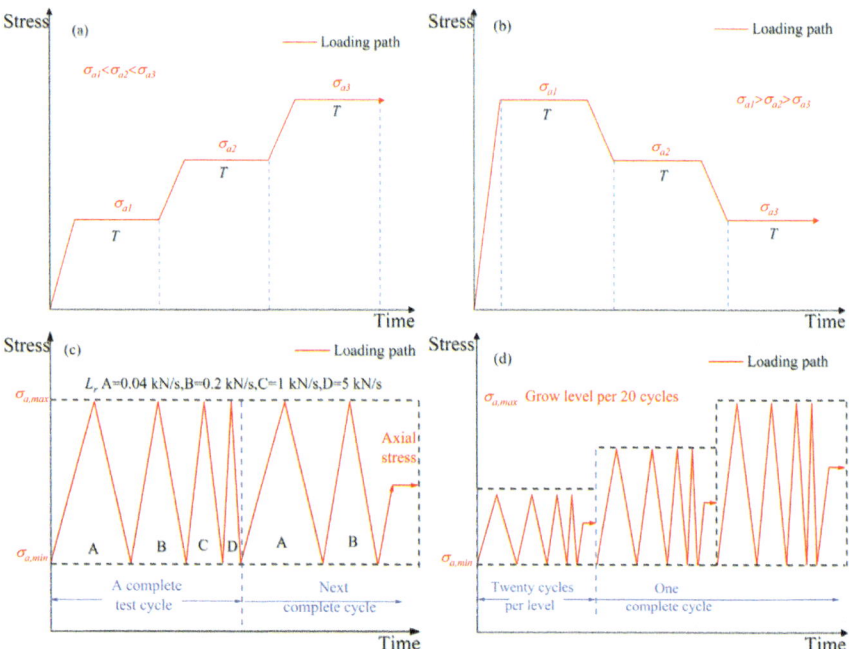

Fig. 3.1 Loading and unloading paths **a** single stress rate fatigue test, **b** fatigue test with different rate mixture and **c** graded mixed rate fatigue test

3.2 Creep Mechanical Properties of Rock Salt Under Different Stress Levels

Table 3.1 Parameters taken in the fatigue tests

Test plan	Upper stress limit/MPa	Lower stress limit/MPa	Loading rate/(kN/s)
ICT	4,8,12,16,20,24	1	1
DCT	20,16,12,8,4	1	1
DMT	25.5	1	0.04, 0.2, 1, 5
GMT	12, 15, 18, 21, 24, 27	1	0.04, 0.2, 1, 5

unloading, maintaining the lower stress limit unchanged, the upper stress limit is increased by 10%, i.e., 6 MPa, until the specimen fails, as shown in Fig. 3.1d.

The specific upper and lower loading and unloading limits for the four sets of tests are shown in Table 3.1. Each group of tests was repeated at least once and valid test data were taken for analysis.

3.2 Creep Mechanical Properties of Rock Salt Under Different Stress Levels

3.2.1 Stress–Strain Curves of Rock Salt Under Step-Up/Step-Down Stress Levels in Creep Tests

The relationship between strain and time for rock salt under incremental stress creep test is shown in Fig. 3.2. The test went through 6 stress levels, with final failure occurring at 24 MPa. From the creep test curve, it can be seen that at each stress level during the instantaneous loading period, the rock salt specimen exhibited a clear instantaneous elastic response, followed by creep deformation that gradually increased over time. In the first four creep stages, the test showed clear decelerating creep stages, also known as attenuating creep stages. At the initial loading stage of each stress level, the axial strain increased with time, while the axial creep rate decreased with time. After a certain period, the creep rate gradually exceeded the steady state, entering a relatively steady creep stage. At the fifth stress level, the creep deformation began to show a third stage, namely the accelerating creep stage, but it was not very pronounced. After entering the sixth stress level, the strain rate of the specimen increased rapidly, entering a clear non-linear accelerating creep stage, exhibiting characteristics of accelerating creep, until the specimen failed. Compared with the axial strain, the radial deformation of the specimen showed a similar trend of change.

The relationship between strain and time for rock salt under decremental stress creep test is shown in Fig. 3.3. During the initial loading process to 20 MPa, the rock salt experienced a large instantaneous strain, followed by a process of strain recovery as the stress decreased. During the subsequent 2 h creep process, the axial strain of the specimen continued to increase, but the creep strain rate began to show a

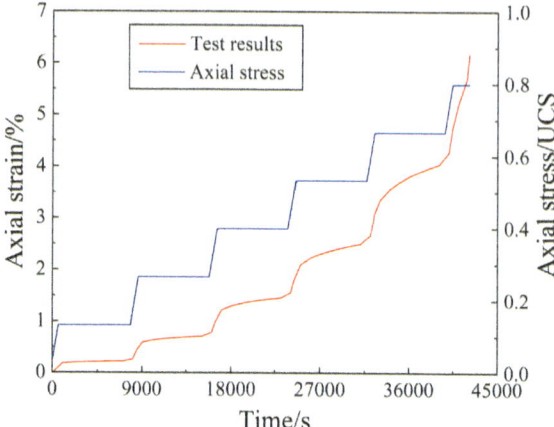

Fig. 3.2 Trends in ICT stress loading paths and axial strain with time

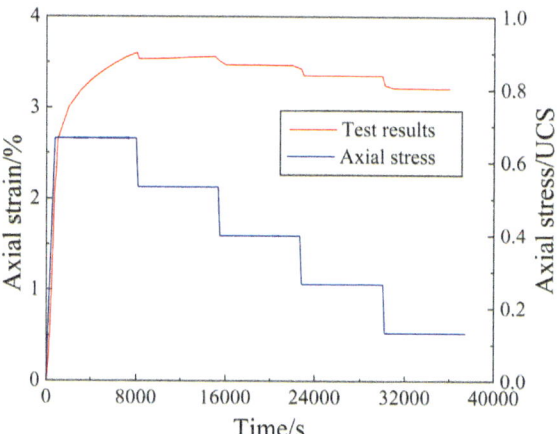

Fig. 3.3 Trends in DCT stress loading paths and axial strain with time

decreasing trend. This indicates that the specimen was in a decelerating creep stage at this time. As the stress level decreased, both the creep rate and creep amount of the rock salt gradually decreased.

3.2.2 The Effect of Stress Level Changes on the Creep Mechanical Properties of Rock Salt

The deformation and deformation rate of the creep variable and the creep variable under decreasing stress were calculated separately to observe the effect of stress level changes on the creep characteristics of rock salt. The calculation results are shown in Table 3.2.

3.2 Creep Mechanical Properties of Rock Salt Under Different Stress Levels

Table 3.2 Comparison of test results

Load/kN	Creep variable and creep rate for increasing stress levels		Creep variable and creep rate for decreasing stress levels	
	Creep variable/%	Creep rate/h	Creep variable/%	Creep rate/h
8	0.057	0.0285	− 0.012	− 0.006
16	0.180	0.090	− 0.002	− 0.001
24	0.304	0.152	− 0.003	− 0.0015
32	0.460	0.230	0.029	0.0145
40	0.930	0.465	1.202	0.601

From Table 3.2, it can be seen that in the creep test with increasing stress levels, as the stress level increases, both the creep rate and creep variable of the rock salt increase significantly. This phenomenon is similar to results obtained in many literature sources. In the creep test with decreasing stress levels, as the stress level decreases, the creep rate and creep variable of the rock salt gradually decrease. However, when the load is less than 16 MPa, a "negative creep" phenomenon appears, which is not observed in conventional step-by-step loading tests. Comparing the creep experiments of decreasing and increasing stress levels, except for the 20 MPa stress level, at other stress levels, the creep deformation rate of rock salt in the decreasing stress test is much smaller than that in the increasing stress test at the same stress level. This is mainly because in the step-by-step loading test, the rock salt is subjected to gradual loading, resulting in internal structure hardening (the hardening process and position density are closely related, which is not discussed here); while in the step-by-step unloading test, 20 MPa is directly loaded to the corresponding stress level, without experiencing the long-term hardening effect similar to the step-by-step loading test, causing the creep deformation rate at the highest stress level in the step-by-step unloading test to be greater than the creep rate in the step-by-step loading test. In the step-by-step unloading test, after the hardening effect of high stress, the creep rate at subsequent stress levels is smaller than that of the step-by-step loading test, because the hardening effect under high stress is much greater than the softening effect produced under low stress.

3.3 Fatigue Mechanical Properties of Rock Salt Under Different Loading and Unloading Rates

3.3.1 The Effect of Mixed-Rate Loading on the Stress–Strain Curve and Residual Strain of Rock Salt

In the actual operation process of salt cavern compressed air energy storage reservoirs, due to the need for peak shaving, the surrounding rock of the reservoir is subjected to complex loads. Most of the time, the rates of gas injection and extraction are not the same. Completing mixed-rate tests on the same specimen can not only analyze the effects of different loading rates on the specimen, but also avoid errors in test results caused by specimen discreteness. From the stress–strain curve of the mixed-rate test (Fig. 3.4), it can be seen that the overall "sparse-dense-sparse" characteristic of the curve has not changed with the continuous change in rate. However, a more obvious periodic characteristic appears internally, which is related to the distribution of loading rates.

The development pattern of residual strain in the mixed-rate test is shown in Fig. 3.5. It can be observed that the development of residual strain, in addition to the conventional three-stage development pattern, also exhibits significant grouping characteristics. Within each group, the residual strain decreases as the corresponding cyclic rate increases. This pattern is consistent with the conclusions from existing rate fatigue tests, namely, the higher the rate, the smaller the residual strain. Here, this phenomenon of residual strain changing with rate is defined as the rate effect.

The variation of the average residual strain in the second stage of the mixed-rate fatigue test with the stress loading rate is shown in Fig. 3.6. It can be observed that the average residual strain in the mixed-rate test tends to decrease as the stress loading rate increases. However, the rate of decrease is slowing down. The possible reasons are as follows: As the loading and unloading rates continue to increase, the specimen

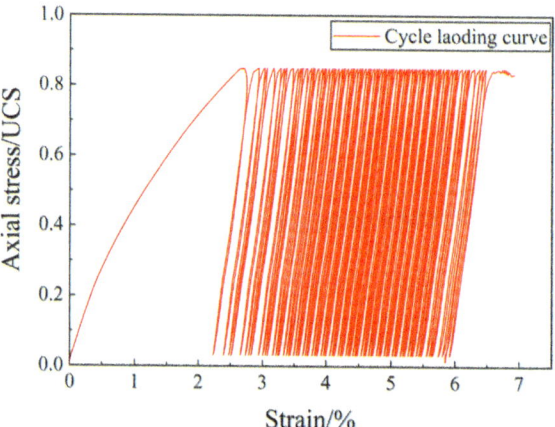

Fig. 3.4 Mixed rate fatigue test stress–strain curve

3.3 Fatigue Mechanical Properties of Rock Salt Under Different Loading ...

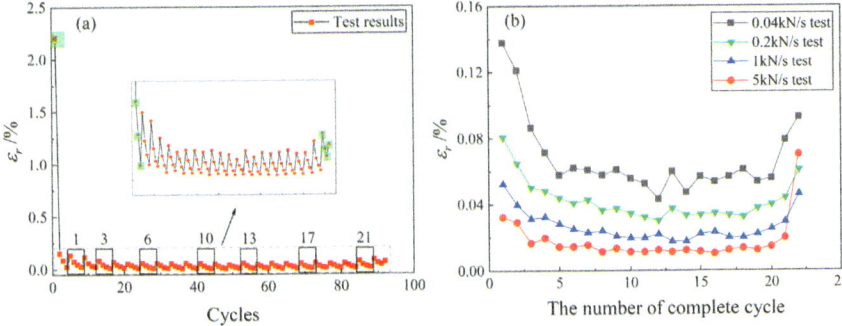

Fig. 3.5 **a** Residual strain curve, **b** residual strain development curves corresponding to cycles at different rates

Fig. 3.6 Average residual strain in the second stage of mixed-rate fatigue test

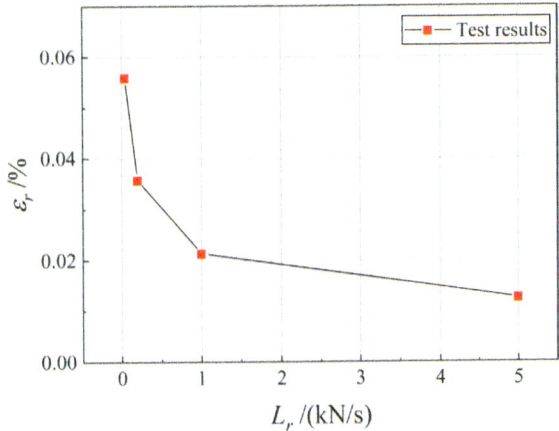

transitions from static loading to impact dynamic loading, and the influence of the loading rate becomes less significant than the impact of the shock on the deformation of rock salt.

3.3.2 The Impact of Stress Levels on the Stress–Strain Curve and Residual Strain of Rock Salt

Another group of graded mixed-rate fatigue tests, where the upper stress level is increased every twenty cycles. The purpose of increasing the upper stress level at equal intervals is to explore the effect of different upper stress limits on the rate effect.

The stress–strain curve for this group of tests is shown in Fig. 3.7. The rock salt sample underwent a total of 65 cycles and failed at the fourth stress level (24 MPa).

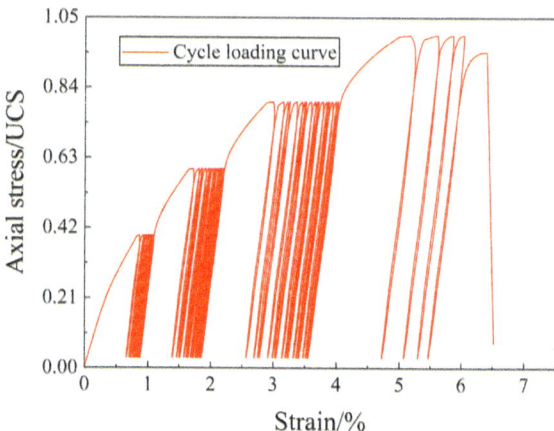

Fig. 3.7 Graded mixed rate fatigue test

The final failure stress was lower than the upper stress limit, which is similar to the experimental results obtained by other researchers.

The development of residual strain is shown in Fig. 3.8a. Within the same stress level, the pattern of residual strain remains consistent with the conclusions from the previous constant-rate fatigue tests. This proves that increasing the upper stress level does not fundamentally change the rate effect of rock salt within the same stress level, i.e., within the same stress level, the higher the rate, the smaller the residual strain produced per cycle. Grouping the residual strain according to the method of the previous set of tests, the residual strains corresponding to different rates within each group are listed, as shown in Fig. 3.9b. It can be clearly seen that as the stress level rises, the cumulative value of residual strain in each group increases significantly. This indicates that the increase in stress level accelerated the development of specimen damage. The reason may be that higher upper stress limits accelerated the development of cracks.

Within different stress level intervals, the mean residual strain for each stress rate corresponding to the cycle was calculated, and the results are shown in Fig. 3.9c. The four curves show a clear upward trend, and the differences between the four curves are also increasing, indicating that the rate effect of rock salt becomes more pronounced as the stress level increases. By taking the difference from largest to smallest within the same stress level interval for the four curves, in the second stress level interval, the average residual strain difference increased by 1.7, 1.05, and 0.41 times, respectively, and in the third stress level interval, it increased by 1.81, 1.3, and 0.79 times, respectively.

By taking every four residual strains as a group and summing them up, as shown in Fig. 3.9, without considering the grouping at the stress level jump, in the second and third stress level intervals, the residual strain increased by about 33% and 30%, respectively. However, when comparing horizontally within different stress levels, the law of the rate effect changes. As shown in Fig. 3.8c, within the second stress level interval, the mean residual strain corresponding to the cycles of 0.02 kN/s, 1

3.3 Fatigue Mechanical Properties of Rock Salt Under Different Loading …   63

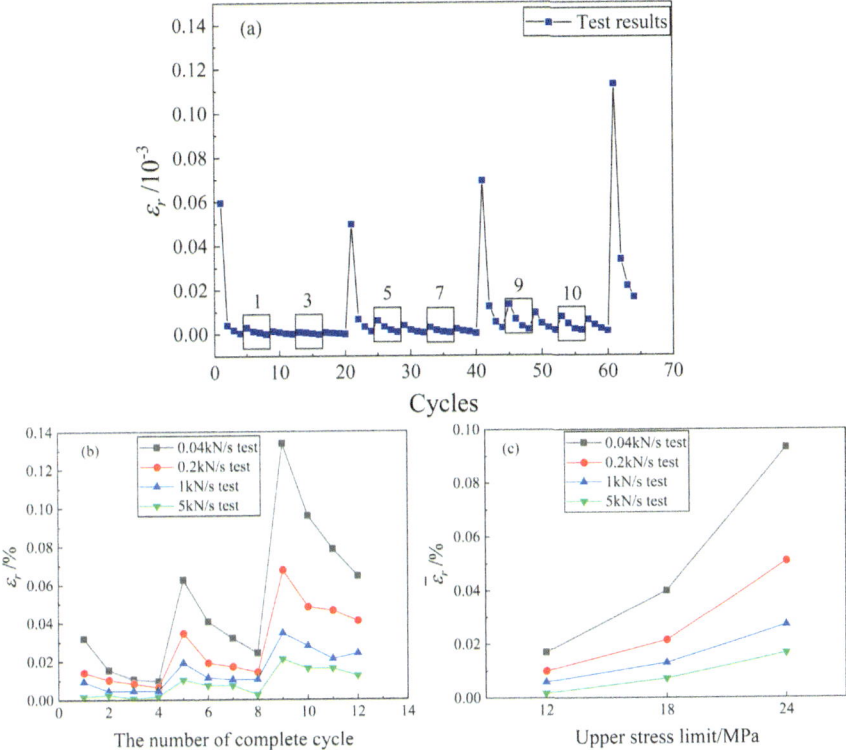

Fig. 3.8 **a** Residual strain development; **b** residual strain mean value of every group; **c** mean value of residual strain in four kinds of velocity

Fig. 3.9 Summation of residual strain in each group

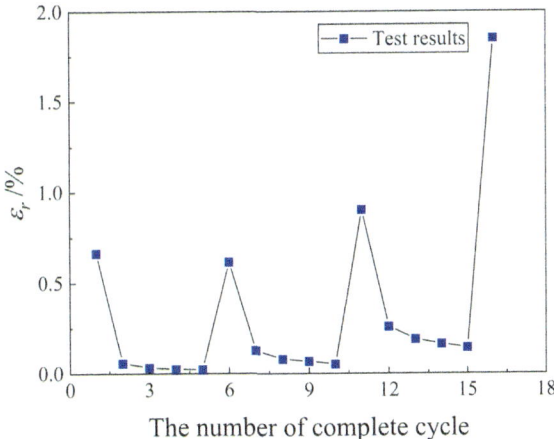

kN/s, and 5 kN/s has already exceeded the mean residual strain corresponding to the cycles of 0.04 kN/s, 0.2 kN/s, and 1 kN/s within the first stress level interval, and the same rule applies between the second and third levels. Moreover, the magnitude of the exceedance accelerates with the increase in stress level, indicating that the influence of stress levels on the fatigue residual strain of rock salt specimens is much greater than the influence brought about by changes in rate, which is noteworthy in actual working conditions.

3.3.3 Quantitative Relationship Between Mixed-Rate Loading and the Mechanical Properties of Rock Salt

To quantitatively analyze the relationship between the development of residual strain under different stress rates on the same specimen, the residual strain development curves from the mixed rate fatigue test shown in Fig. 3.5a were used. In these curves, every four continuous cycles with different stress rates were taken as a group, and the function $y = ax^{-b}$ was fitted to these groups (as shown in Fig. 3.5a, each small box from 1 to 20 contains four points forming a group; here, a is defined as the deformation coefficient of the specimen, 'b' is the correlation parameter related to the loading stress rate, and x is the value of the rate, which is always greater than 0). All grouped curves had a good fitting effect, as shown in Fig. 3.10a, which displays some of the fitted curves, while Table 3.3 provides partial fitting parameter results. Fig. 3.10b and c illustrate the development of the a and b parameters from the fitting results. It can be observed that the fitting curve coefficient a exhibits a flattened U shaped trend, which not only conforms to the three-stage deformation pattern of rock salt fatigue but also aligns with the deformation pattern of grouped residual strains in the mixed rate test shown in Fig. 3.5b. This indicates that the fitted a value can effectively represent the deformation capacity of the specimen throughout the entire fatigue testing process.

In the power function $y = ax^{-b}$, the magnitude of the exponent affects the overall position of the function's graph. The smaller the value of b, the flatter and smoother the graph, which is expressed as the value of y in the equation strictly monotonically decreasing. Therefore, $b > 0$ always holds true. Another obvious characteristic is that the exponent 'b' always fluctuates within a certain range. This indicates that the correlation parameter related to the loading stress rate, as an inherent property of the material itself, may not change regularly with the change of the loading rate itself but fluctuates around the average value. This provides a guiding direction for the proposal of the rock salt rate equation in the next phase.

Similarly, in the mixed rate fatigue tests with graded stress limits, within each stress level interval, the residual strain values corresponding to four consecutive cycles with different rates are grouped together. For the initial first cycle and the cycle with a stress level jump, due to the large difference in deformation, they are not grouped with the next three adjacent cycles for data fitting. Therefore, only four

3.3 Fatigue Mechanical Properties of Rock Salt Under Different Loading ...

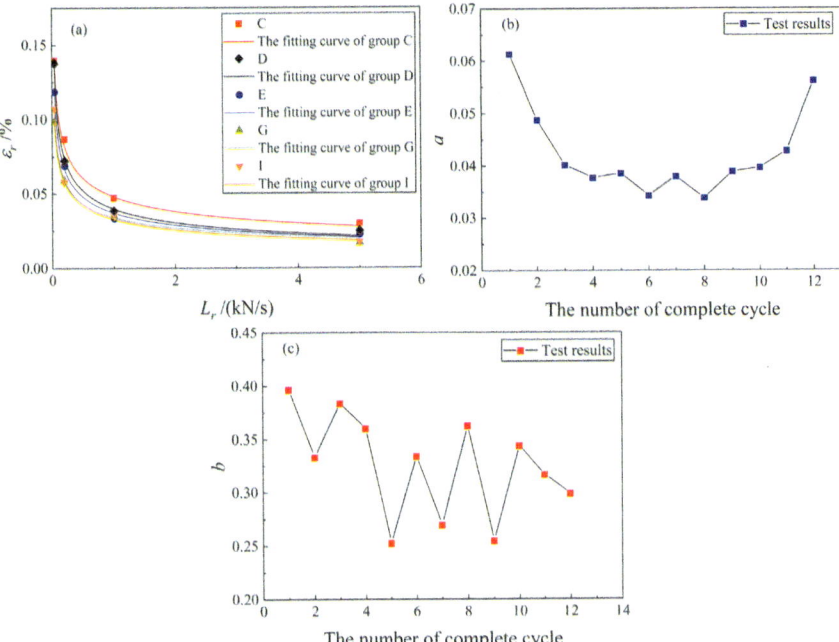

Fig. 3.10 **a** Partial fitting curves (group C, D, E, G, I); **b** development trend of coefficient A; **c** development trend of index B in rate-coupled fatigue test

Table 3.3 Fitting results of all residual strain groups in rate-coupled fatigue test

$y = ax^{-b}$ fitting results			
	a	b	R^2
A	0.06141	0.39658	0.99889
B	0.04880	0.33335	0.99526
C	0.04022	0.38376	0.99801
D	0.03779	0.36037	0.99576
E	0.03859	0.25265	0.96542
F	0.03435	0.33404	0.99636
G	0.03799	0.26956	0.95848
H	0.03382	0.36248	0.99823
I	0.03886	0.25471	0.92506
J	0.03962	0.34380	0.98891
K	0.04277	0.31659	0.99864
L	0.05616	0.29867	0.99883

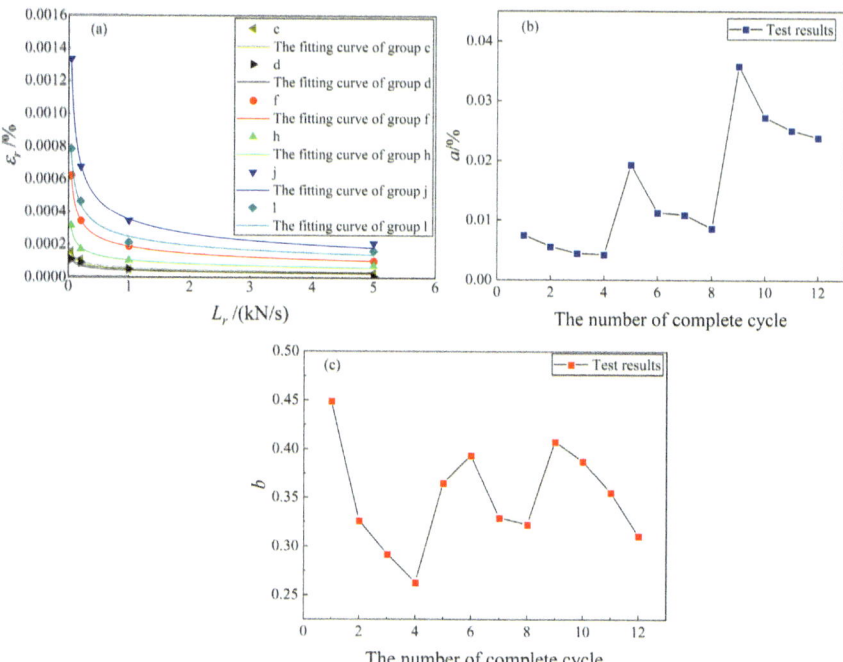

Fig. 3.11 **a** Partial fitting curves (group c, d, f, h, j, l); **b** development trend of coefficient A; **c** development trend of index B in rate-coupled hierarchical fatigue test

groups are divided within each stress level, and the residual strain values of the four cycles in each group are fitted with the formula $y = ax^{-b}$. As shown in Fig. 3.11a, some of the fitted curves are displayed, while Fig. 3.11b and c show the development curves of the coefficients 'a' and the exponent 'b', respectively, and Table 3.4 presents the fitting results.

From the developmental pattern of the a values in Fig. 3.11b, a distinct stress grading characteristic can be observed. The trend of these values is positively correlated with the development trend of the cyclic residual strain values statistically analyzed at different rates in Fig. 3.8b. This indicates that in tests where the stress limit levels are altered, the a values can still represent the accumulation pattern of residual strain. In Fig. 3.11c, the 'b' values also exhibit a clear variation with stress levels, and within the same stress level, these values show a distinct downward trend. However, they still exhibit a noticeable interval distribution characteristic, which is consistent with the conclusions drawn from the mixed rate experiments.

Table 3.4 Fitting results of all residual strain groups in rate-coupled hierarchical fatigue test

$y = ax^{-b}$ fitting results

	a * 10^{-4}	b	R^2
a	0.750758	0.44923	0.97362
b	0.556794	0.32651	0.96352
c	0.444573	0.29233	0.82291
d	0.424950	0.26319	0.94349
e	1.936850	0.39399	0.99999
f	1.128260	0.36514	0.98342
g	1.091500	0.32971	0.98602
h	0.865936	0.32283	0.94211
i	3.584950	0.40805	0.99797
j	2.730070	0.38806	0.99552
k	2.510870	0.35579	0.98523
l	2.393560	0.31142	0.99299

3.3.4 Rate Effect Equation of Rock Salt Influenced by Stress Loading Rate

Different loading and unloading rates cause different amounts of residual deformation, also known as plastic deformation, which is referred to here as the rate effect. In elastoplastic models, if the principal stress exceeds the yield surface, deformation will produce new plastic deformation according to the flow rule. During the deformation process, the yield surface and flow rule undergo hardening/softening movement according to the hardening rule. The predictive ability of such elastoplastic models for the mechanical behavior of rock salt is limited because the plastic deformation of rock salt exhibits a strong time dependency (or rate dependency), known as viscoplastic deformation. To characterize the time dependency of plastic deformation, creep models have been proposed to represent plastic deformation related to time. In cyclic loading and unloading tests, there is both time-dependent deformation and time-independent deformation. To differentiate between creep plastic deformation and loading plastic deformation in residual deformation, it is necessary to establish a rate effect equation based on the time dependency of rock salt deformation.

In the previous text, it can be obtained that in the function $y = ax^{-b}$, the value of a can reflect the degree of rock salt deformation ability, while the value of 'b' is related to the deformation properties of the rock salt itself. Although this equation is relatively simple, it can provide a reference for the proposal of the rate effect equation.

To analyze the quantitative relationship of the rate effect and to explore the difference between the time-related creep action and the time-independent loading action under the same stress path in different rate cycles, it is considered that the total plastic deformation in each cycle can be divided into time-related creep plastic deformation and time-independent loading plastic deformation. The creep plastic deformation can be described using a creep constitutive model. The commonly used creep constitutive model is the Norton model:

$$\dot{\varepsilon}_v(t) = A\sigma^n \tag{3.1}$$

In the equaton, $\dot{\varepsilon}_v(t)$ represents the viscous strain rate and σ represents the stress. Both A and n are material parameters. This model is usually used to determine the steady-state creep rate. However, during the loading and unloading process, the stress is constantly changing. With each entry into a new stress state, the deformation of rock salt actually develops from instantaneous creep to steady-state creep. Therefore, a creep constitutive model that describes the entire process (at least including the first two stages) is needed. The Burgers nonlinear creep model constructed by Wang Junbao et al. and the fractional order nonlinear creep constitutive equation constructed by Zhou Hongwei et al. have good predictive effects on creep deformation, as shown in Eqs. (3.2) and (3.3), respectively:

$$\dot{\varepsilon}_v(t) = \frac{\sigma^n}{\eta}\left(1 + \frac{c}{2\sqrt{t}}\right) \tag{3.2}$$

$$\dot{\varepsilon}_v(t) = \beta \frac{\sigma}{\eta} \frac{t^{\beta-1}}{\Gamma(\beta+1)} \tag{3.3}$$

In the equation, η, b and c are all material parameters, and $\Gamma(\beta+1)$ is the Gamma function. It can be observed that the creep plastic deformation rate is a function of both stress and time, all expressed as a combination of power functions of stress and time. For convenience in derivation, this paper also uses a combination of stress and time power functions to describe the time-related creep plastic deformation, as shown in Eq. (3.4):

$$\varepsilon_v = a\sigma^n * t^b \tag{3.4}$$

In the equation, a, n, and b are all material parameters. To exclude the influence of loading plasticity, a set of constant load creep test data ($\sigma = 21 MPa$) was used for verification, and it was found that the fitted relationship between time and deformation had a good effect, as shown in Fig. 3.12.

Take the first derivative of Eq. (3.4) to obtain the creep plastic deformation rate, as shown in Eq. (3.5).

$$\dot{\varepsilon}_v = \sigma^n ab * t^{b-1} \tag{3.5}$$

In the creep constitutive model of rock salt, the typical value of n ranges between 1 and 2, such as n = 1 in the fractional order model. The total plastic strain in the cyclic process can be expressed as:

3.3 Fatigue Mechanical Properties of Rock Salt Under Different Loading ...

Fig. 3.12 Creep test fitting curve

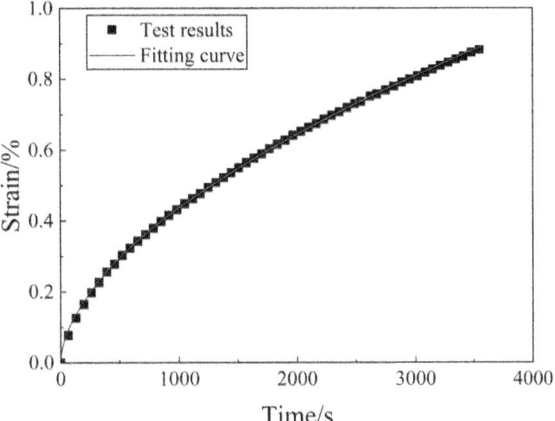

$$\varepsilon_p = \varepsilon_v + \varepsilon_l \qquad (3.6)$$

Integrate with respect to time:

$$\varepsilon_p = \int_0^t \dot{\varepsilon}_v dt + \varepsilon_l \qquad (3.7)$$

Substitute Eq. (3.5) into Eq. (3.7) to get:

$$\varepsilon_p = \int_0^t \left(\sigma^n ab * t^{b-1}\right) dt + \varepsilon_l \qquad (3.8)$$

For convenience in derivation, we temporarily take n = 1 here. We also derived the formula for n = 2 and found that the value of 'n' does not affect the final formula result. Equation (3.8) can be transformed into the form of Eq. (3.9):

$$\varepsilon_p = \int_0^t \left(\sigma ab * t^{b-1}\right) dt + \varepsilon_l \qquad (3.9)$$

During the loading and unloading process, the stress is in the form of a constant stress rate, $\sigma = \sigma_0 + vt$, where σ_0 is the initial stress and v is the loading rate. Substitute Eq. (3.9) to transform it into the relationship between loading time and acceleration rate:

$$\varepsilon_p = \int_0^t (\sigma_0 + vt)(ab * t^{b-1}) dt + \varepsilon_l \qquad (3.10)$$

Integrate the cyclic loading section from time 0 to t:

$$\varepsilon_p = \left(\sigma_0 a * t^b + \frac{vab}{b+1} t^{b+1} \right) \Big|_0^t + \varepsilon_l \qquad (3.11)$$

Assume a test as the baseline test, with its loading rate v_0, then the loading time is t_0. Since the upper limit stress is the same during loading, the total plastic deformation during the loading phase at different loading rates can be expressed as:

$$\varepsilon_p = \sigma_0 a \left(\frac{v_0}{v} t_0 \right)^b + \frac{vab}{b+1} \left(\frac{v_0}{v} t_0 \right)^{b+1} + \varepsilon_l \qquad (3.12)$$

After organizing and combining like terms, we get:

$$\varepsilon_p = v^{-b} \left[\sigma_0 a (v_0 t_0)^b + \frac{ab}{b+1} (v_0 t_0)^{b+1} \right] + \varepsilon_l \qquad (3.13)$$

Since the upper and lower limits of the loading stress are the same, and the material parameters are also the same in the tests conducted on the same specimen, only the loading rate is the independent variable in Eq. (3.13).

Make $\sigma_0 a(v_0 t_0)^b + \frac{ab}{b+1}(v_0 * t_0)^{b+1} = m$, then we have:

$$\varepsilon_p = m v^{-b} + \varepsilon_l \qquad (3.14)$$

Equation (3.14) is the relationship equation between the overall plasticity and the rate in the cyclic loading and unloading process. It is worth noting that although the derivation process is aimed at the loading phase, the formula also applies to the unloading phase because the upper and lower limits of the unloading rate are the same. That is, the residual deformation of the entire loading and unloading cycle process satisfies the relationship equation.

Parameter m represents the creep stress effect and can be called the stress factor, and b represents the effect of the loading rate and can be called the rate factor.

3.3.5 Deformation Analysis of Rock Salt During Stress Loading Process

The residual deformation in the rate-coupled cyclic loading and unloading test is fitted using Eq. (3.14), and the variation of m, b, and ε_l is shown in Fig. 3.13. In Fig. 3.13a, parameter m (i.e., the stress factor) reflects the degree of rock salt creep

3.3 Fatigue Mechanical Properties of Rock Salt Under Different Loading ...   71

ability throughout the process, showing an overall trend of decreasing, stabilizing, and then increasing in a U shape. Initially, due to the loose internal structure of the specimen, and near the point of failure due to the rapid expansion of cracks, the deformation is larger, resulting in higher 'm' values in the early and late stages of the test.

In Fig. 3.13b, the parameter b (i.e., the creep rate factor) reflects the variation in the creep rate, which fluctuates around $0.41 \pm 0.01 \sim 0.03$ with a small amplitude. This is mainly because the rate correlation is an inherent property of the material itself, maintaining a constant value.

Based on Eq. (3.6), the residual deformation is divided into creep plastic deformation (ε_v) and loading plastic deformation (ε_l). Figure 3.13c shows the development process of plastic deformation in each cycle of the rate-coupled test. From the parameter values of the fitting results, it can also be seen that in the second stage where the creep deformation ability is stable, the creep plastic deformation is much greater than the loading plastic deformation. However, as the loading and unloading rate increases from 0.04 to 5 kN/s, the amount of creep plastic deformation decreases rapidly. The

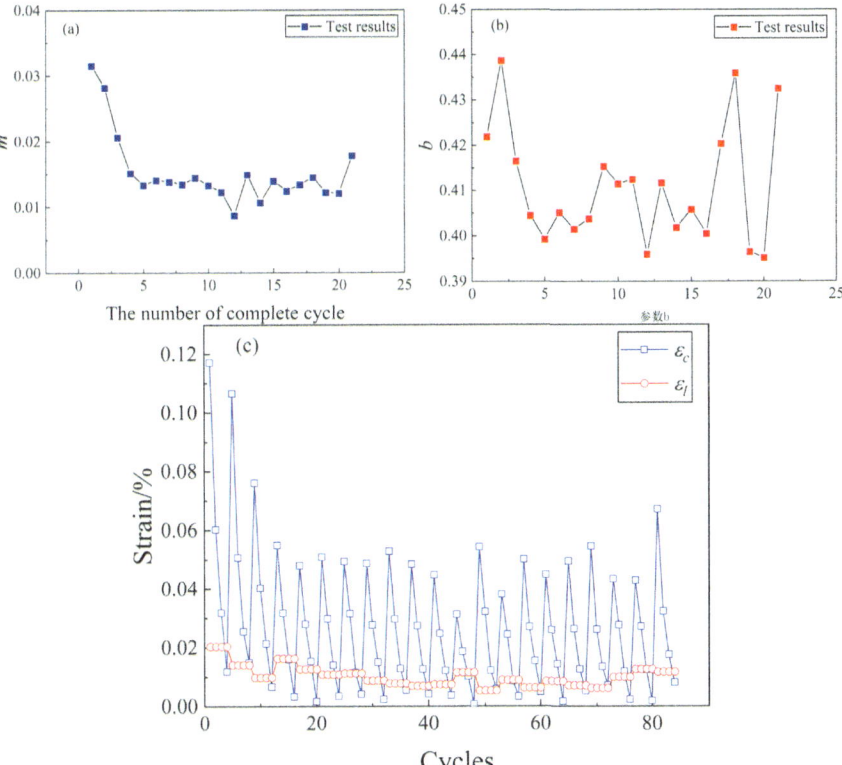

Fig. 3.13 **a** Development of parameter m; **b** distribution of parameter b; **c** Distribution of creep strain and fatigue strain calculated by Eq. (3.14)

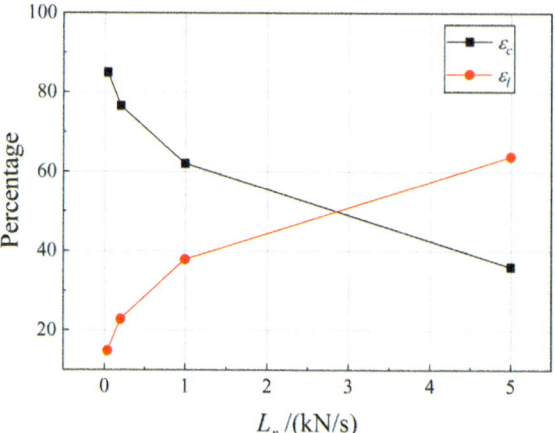

Fig. 3.14 Average proportion of creep plastic deformation and loading plastic deformation in different rate cycles

average values of creep plastic deformation per cycle in the overall second stage are 4.772×10^{-4}, 2.371×10^{-4}, 1.281×10^{-4}, and 0.394×10^{-4}, respectively.

As the loading and unloading rate increases, the proportion of creep plastic deformation decreases exponentially, from 85 to 36% (as shown in Fig. 3.14, which are 85%, 77%, 62%, and 36%, respectively). This indicates that as the cyclic loading and unloading rate increases, the creep effect during the loading process of rock salt gradually decreases, while the loading plastic deformation, which is independent of time, gradually becomes dominant, increasing from 15 to 64%. This also illustrates that the fatigue effect is much greater than the creep effect during rapid loading and unloading processes.

3.4 Conclusions

Due to the need for peak shaving, compressed air energy storage power stations will change the magnitude and frequency of gas injection and extraction, which causes significant changes in the severity of the load borne by the surrounding rock of the salt cavern storage, that is, different levels and rates of stress loading. Against this backdrop, this chapter has studied the creep and fatigue mechanical properties of rock salt. It compared and analyzed the impact of increasing/decreasing stress levels on the creep deformation of rock salt and conducted a detailed study of the impact of stress loading rates on the fatigue of rock salt. Based on the Norton constitutive model, a rate effect equation for rock salt was established, and the plastic characteristics during the loading and unloading process, namely creep plastic deformation and loading plastic deformation, were analyzed on this basis. The main conclusions are as follows:

(1) Creep experiments on rock salt with different loading and unloading paths were carried out, and it was found that the loading and unloading history has a significant impact on the creep behavior of rock salt. In the incremental stress graded creep experiments, the creep rate and creep amount of rock salt increased with the increase of stress levels. In the decremental stress graded creep experiments, except for the highest stress level, the creep deformation rate of rock salt was much smaller than the creep rate in the same stress level of the incremental stress test. At low stress levels in the decremental stress creep experiment, zero creep or negative creep phenomena occurred.

(2) Mixed rate fatigue tests on rock salt under different stress rates were conducted. It was found that the loading rate significantly affects the fatigue mechanical properties of rock salt, with faster loading rates resulting in smaller residual strains per cycle. The fatigue of rock salt exhibits a clear rate effect. An increase in stress levels amplifies the rate effect on rock salt.

(3) Based on the Norton constitutive model, a rate effect equation was established by integrating the creep deformation at a fixed rate over time. In the rate effect equation, parameter m represents the degree of creep ability, and parameter b represents the change in creep rate. Based on the rate effect equation, the deformation during the fatigue process of rock salt is divided into time-related deformation (creep plastic deformation) and time-unrelated deformation (loading plastic deformation). The creep plastic deformation decreases with the increase of loading and unloading rates, while the loading plastic deformation first decreases and then increases with the fatigue process.

References

1. Luxuan Tang, Jinyang Fan, Zongze Li, et al. A new constitutive model for salt rock under cyclic loadings based on state variables[J]. Geoenergy Science and Engineering, 2024, 233: 212433.
2. Zhenyu Yang, Jinyang Fan, Jie Chen, et al. Dilatancy and Acoustic Emission Characteristics of Rock Salt in Variable-Frequency Fatigue Tests[J]. Rock Mechanics and Rock Engineering, 2024: 1–18.
3. Wenhao Liu, Deyi Jiang, Jinyang Fan, et al. Experimental study on effect of cyclic gas pressure on mechanical and acoustic emission characteristics of salt rock[J]. Journal of Energy Storage, 2024, 99: 113410.
4. Xuan Wang, Hongling Ma, Hang Li, et al. Projected effective energy stored of Zhangshu salt cavern per day in CAES in 2060[J]. Energy, 2024, 299: 131283.
5. Junbao Wang, Qiang Zhang, Zhanping Song, et al. Nonlinear creep model of salt rock used for displacement prediction of salt cavern gas storage[J]. Journal of Energy Storage, 2022, 48: 103951.

Open Access This chapter is licensed under the terms of the Creative Commons Attribution-NonCommercial-NoDerivatives 4.0 International License (http://creativecommons.org/licenses/by-nc-nd/4.0/), which permits any noncommercial use, sharing, distribution and reproduction in any medium or format, as long as you give appropriate credit to the original author(s) and the source, provide a link to the Creative Commons license and indicate if you modified the licensed material. You do not have permission under this license to share adapted material derived from this chapter or parts of it.

The images or other third party material in this chapter are included in the chapter's Creative Commons license, unless indicated otherwise in a credit line to the material. If material is not included in the chapter's Creative Commons license and your intended use is not permitted by statutory regulation or exceeds the permitted use, you will need to obtain permission directly from the copyright holder.

Chapter 4
Discontinuous Fatigue Mechanical Properties for Rock Salt

The previous chapter introduced the conventional fatigue and properties for rock salt. Conventional fatigue and creep means the material is subjected to continuous loading [2]. However, in reality geomaterials the loading typically is regular, but randomly interrupted a reactivated [3]. For instance, Germany Huntorf plant completes one cycle every day [1]. The underground salt CAES works 10 h and stop 14 h. During the period of stop, the air pressure inside CAES is constant. Whether this pressure plateau would exert influence on the fatigue properties is unclear [4].

4.1 Experimental Methods

To ensure the comparability of experimental result, the same samples and equipment are used in the following discontinuous cyclic loading tests. The rock salt samples have standard cylindrical shape of 50 mm × 100 mm.

A set of continuous fatigue tests (CFT) and four sets of discontinuous fatigue tests (DFTs) were conducted. The upper stress limit for the tests was set at 85% of the UCS, while the lower stress limit was set at 3% of the UCS, with a loading and unloading rate of 2 kN/s. The low-stress interval times for the discontinuous cycles were set to 5, 10, 15, and 20 minutes, respectively. The specific loading and unloading paths are shown in Fig. 4.1. The cycles before the low-stress interval are defined as A cycles, and the cycles after the interval are defined as B cycles. The first cycle (F), which does not belong to either the pre-interval or post-interval cycles, is defined as the initial cycle.

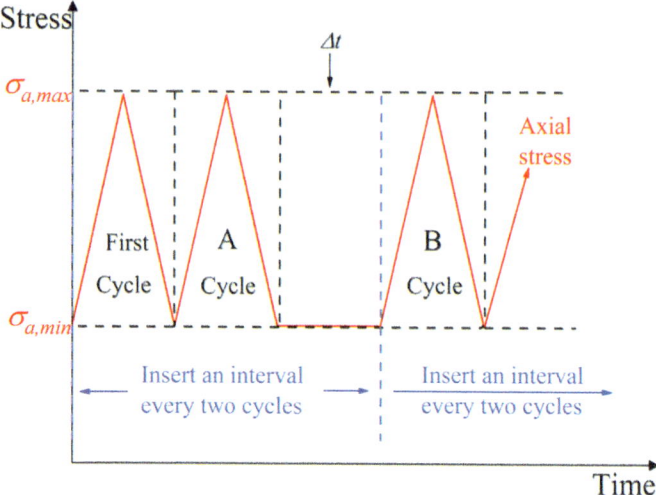

Fig. 4.1 Loading path for CFT and DFTs

4.2 Test Results and Analysis

4.2.1 Stress–Strain Curves

Discontinuous fatigue tests show a distinct difference from the conventional tests. As shown in Fig. 4.2, the fatigue life of the salt samples from CFT and DFT are 89, 34, 24, 13, 20, respectively. The fatigue live of salt from DFTs is significantly lower than that from the conventional one under the same conditions.

Calculating the total accumulated plastic deformation (except the last uncomplete cycle), this accumulated plastic deformation of the salt samples from DFTs is 7.2–8.5%, smaller that from conventional fatigue tests, 9.1–11%.

4.2.2 Residual Strains During in the Rock Salt Test

Every loading generate a certain deformation. While one part of it (elastic deformation) is recovered during the unloading, the other, is not. The remaining part is plastic strain, also called the residual strain (ε_r).

Considering that A cycle and B cycle of the creep–fatigue test are always in different positions (the A cycle is always in the even position, and the B cycle is always in the odd position, including the first cycle), the ε_r of the A and B cycles at the corresponding positions were solved by using the arithmetic mean method, as shown in the following equations:

4.2 Test Results and Analysis

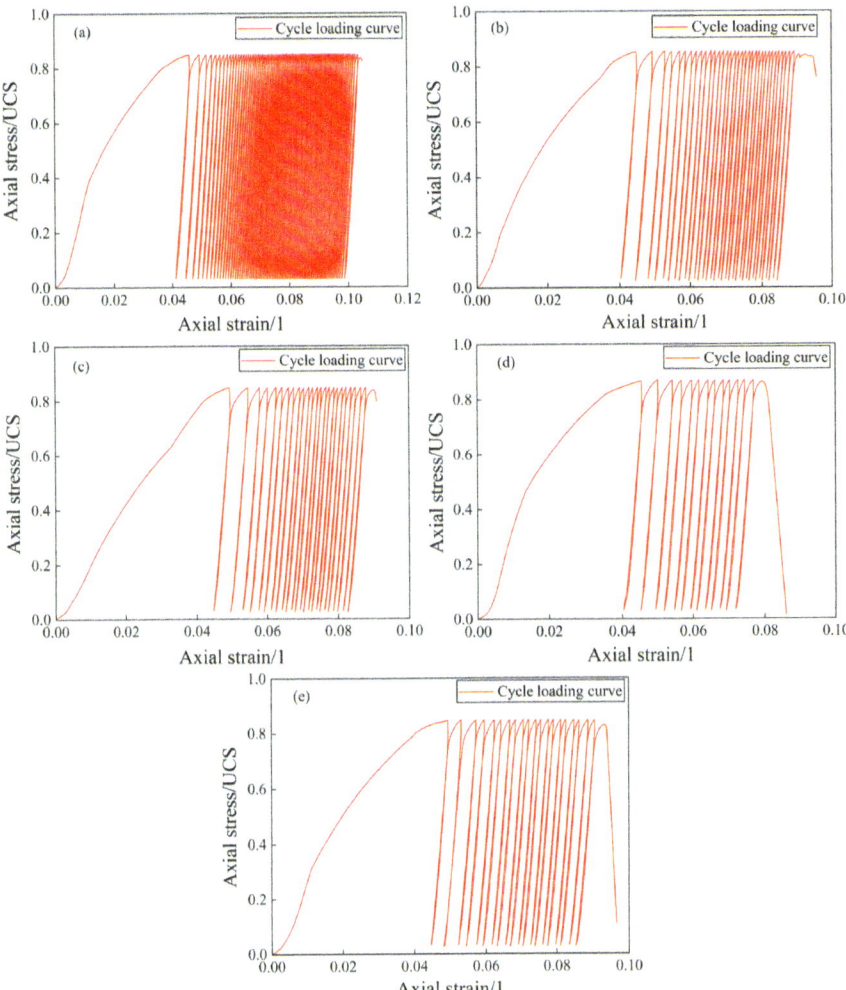

Fig. 4.2 Axial stress-axial strain plot from CFT and DFTs, **a** conventional fatigue test, **b** discontinuous fatigue test ($\Delta t = 5$ min), **c** discontinuous fatigue test ($\Delta t = 10$ min), **d** discontinuous fatigue test ($\Delta t = 15$ min) and **e** discontinuous fatigue test ($\Delta t = 20$ min)

$$\varepsilon_{r,Ai} = \left\{ \begin{array}{c} \varepsilon_{r,Ai}, i = even\, 2 \leq i < N \\ \frac{\varepsilon_{r,A(i-1)} + \varepsilon_{r,As(i+1)}}{2}, i = odd, 3 \leq i < N \end{array} \right\} \quad (4.1)$$

$$\varepsilon_{r,Bi} = \left\{ \begin{array}{c} \frac{3\varepsilon_{r,B3} - \varepsilon_{r,B5}}{2}, i = 2 \\ \varepsilon_{r,Bi}, i = odd\, 3 \leq i < N \\ \frac{\varepsilon_{r,B(i-1)} + \varepsilon_{r,B(i+1)}}{2}, i = even, 4 \leq i < N \end{array} \right\} \quad (4.2)$$

where $\varepsilon_{r,Ai}$ and $\varepsilon_{r,Bi}$ are the ε_r results of the A and B cycles in the ith cycle, respectively, and N is the number of loading cycles.

Calculating the residual strain in every cycle leads to the understanding of the evolution of plastic damage of salt sample under discontinuous cyclic loading. Figure 4.3 shows the residual strain evolution with stress cycles. The residual strain firstly reduces, then keeps at a constant and finally increases a little when close to failure. In previous studies, the first stage is called as decelerated deformation phase; the second is uniform deformation phase; the last is accelerated deformation phase [5]. In the aspects of these three phases, DFTs show the same features with the conventional.

4.2.3 Elastic Constants from the Rock Salt Test

The elastic modulus is calculated from the slope of the linear segment BC segment is calculated as in Fig. 4.4. In the conventional fatigue tests, the elastic modulus increases with the stress cycles (Fig. 4.5).

Salt comprises many crystalline grains. The intact grain could orderly release the elastic strain energy. As the fatigue damage accumulates, the number of the crushed (damaged) grains rises and the incremental elastic strain reduces, leading to the elastic modulus increases. For every loading, the slipping path of dislocation is random. In different path, the intact functionary grain may be different, so the elastic modulus fluctuates lightly. Surely, this fluctuation is also related to the error from the measurement device and data processing.

The calculated Poisson's ratio for DFTs is shown in Fig. 4.6. The Poisson's ratio did not show any significant difference between A and B path. As the whole trend, the Poisson's ratio increases with cycles, both for conventional fatigue tests and DFT group tests.

Dilatancy angle reflects the relation between the increment of plastic volumetric deformation and increment of plastic shear deformation. Dilatancy angle is related to the mechanism of plastic deformation: plastic shear strain is due the dislocation slippage before the initiation of microcracking. Volumetric expansion depends on the vacancy and lattice distortion, thus the dilatancy angle (or dilatancy factor) is constant. If microcracks propagate rapidly, the dilatancy is influenced by increase in space of microcracks and the dilatancy angle increases with plastic shear strain.

The average tangent of dilatancy angle in every cycle was calculated. It rises with stress cycle, showing the contribution of microcracks to dilatancy growth. Although the dilatancy angle varies differently between A and B, the developing trends in A path and B path are the same.

4.2 Test Results and Analysis

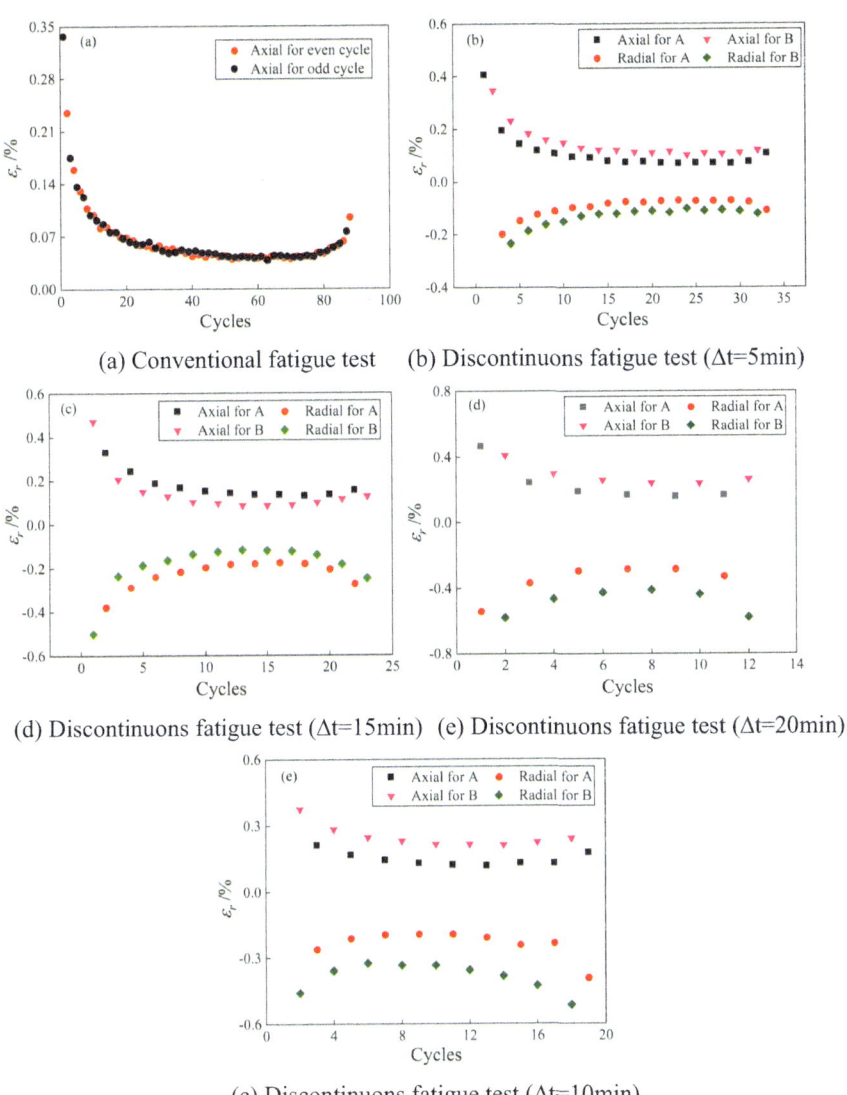

Fig. 4.3 Evolution of the residual strain with loading cycles from CFT and DFTs

4.2.4 Time Interval Effect on Rock Salt

The scattering of experimental results brings much inconvenience to the analysis. To investigate the effect of time interval on the fatigue properties and avoid the discreteness of different samples, the validation discontinuous fatigue test (VDFT) is conducted following the loading path in Fig. 4.7.

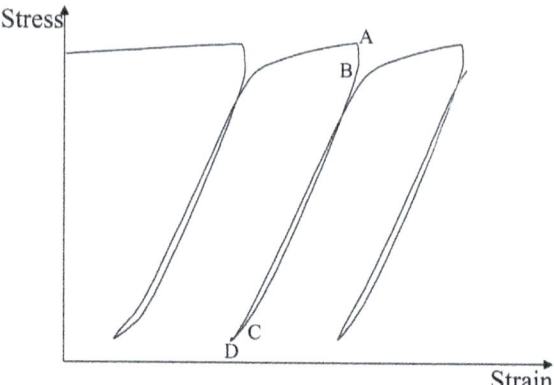

Fig. 4.4 Schematic diagram of the method for calculation of elastic module

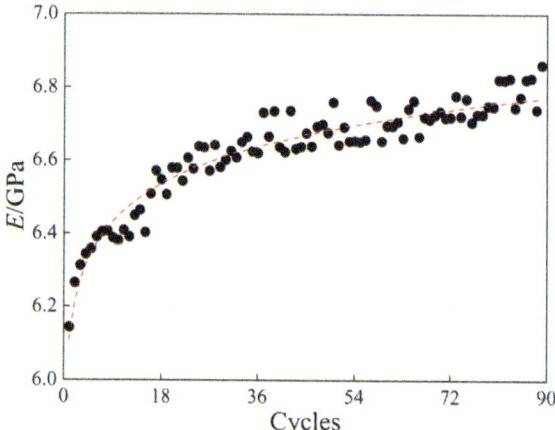

Fig. 4.5 Evolution of elastic modulus in CFT

The stress ratio, loading velocity and temperature applied in VDFT are the same with DFTs. Every stress cycle is followed by an interval. The intervals are 0, 5, 10, 15 and 20 min, in turn. Every five cycles and relevant interval are considered as the complete cycle.

The VDFT combines the interval effect of 0, 5, 10, 15 and 20 min. 38 cycles, 7 mega-cycles were completed. The total accumulated plastic strain is 8.4%, much lower than in the conventional fatigue test. The calculated residual strain are shown in Fig. 4.8.

Except the first complete cycle, the residual strain increases within the complete cycle as the time interval elongates, indicating the time interval could exert notable influence on the discontinuous fatigue properties of rock salt.

4.2 Test Results and Analysis

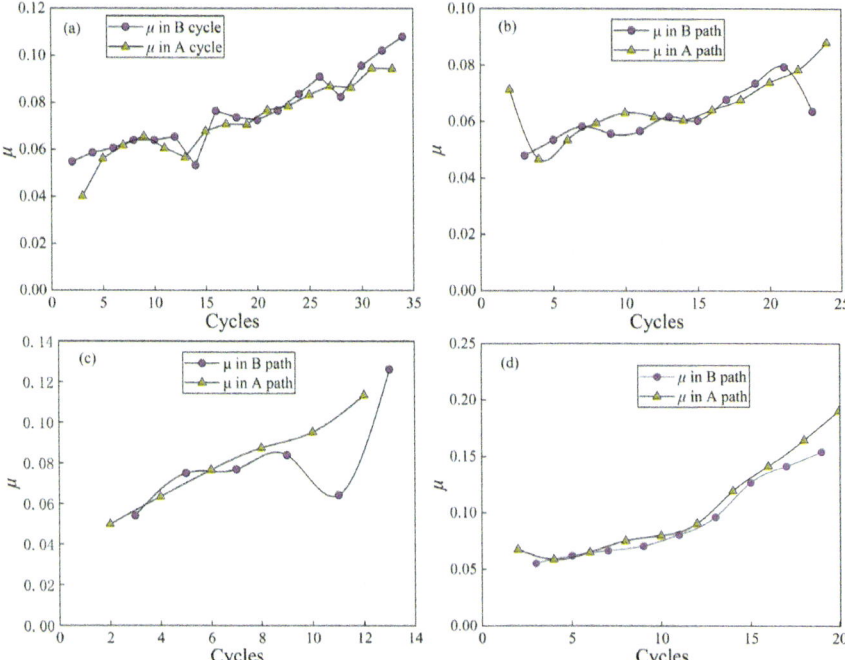

Fig. 4.6 Variation of the Poisson's ratio from the DFTs, **a** discontinuous fatigue test ($\Delta t = 5$ min), **b** discontinuous fatigue test ($\Delta t = 10$ min), **c** discontinuous fatigue test ($\Delta t = 15$ min), **d** discontinuous fatigue test ($\Delta t = 20$ min)

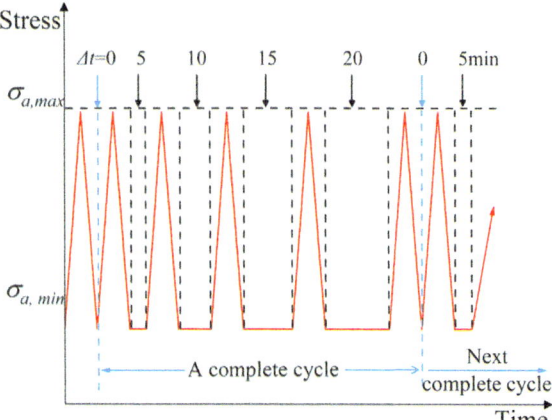

Fig. 4.7 Loading path for VDFT

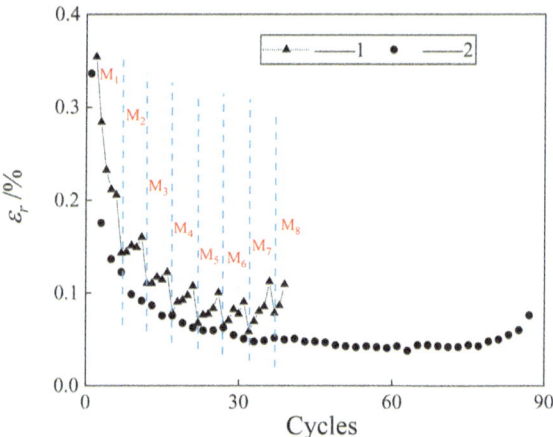

Fig. 4.8 Variation of the residual (axial) strain from the VDFT (indicated with 1) and CFT (indicated with 2)

4.2.5 Rupture Form

The rupture shape of salt from discontinuous fatigue and conventional fatigue is similar for uniaxial compression test (Fig. 4.9). The cracked rock salt samples appear mixed fracturing with split and shear. Two parts form, one fusiform structure inside the sample (A part in Fig. 4.10c) and cylindrical wall outside (B part in Fig. 4.10d). The fusiform structure finally was cut by a thoroughgoing shear crack, divided into two cone parts (as shown in Fig. 4.9c and d).

The fusiform structure forms because of the end friction. When the fusiform structure is compressed and expands around, the B part has to subjected to the squeezing from A part and generates tensile cracks (as shown in Fig. 4.9a and b). CT scanner was applied to scan the intersections of a sample which has complete three cycles

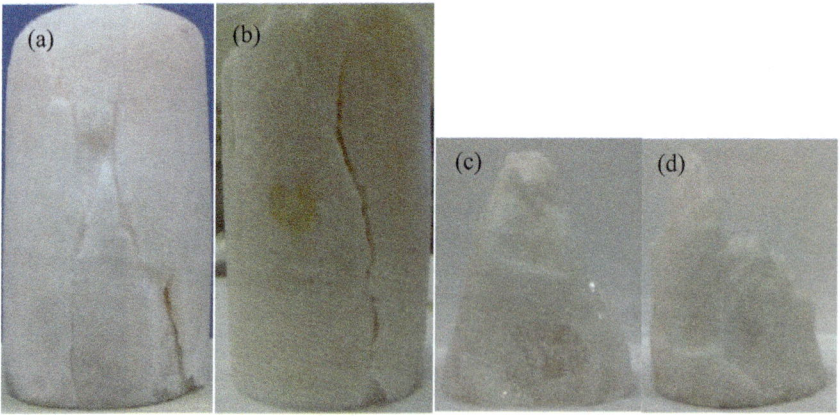

Fig. 4.9 Fractured sample

4.2 Test Results and Analysis

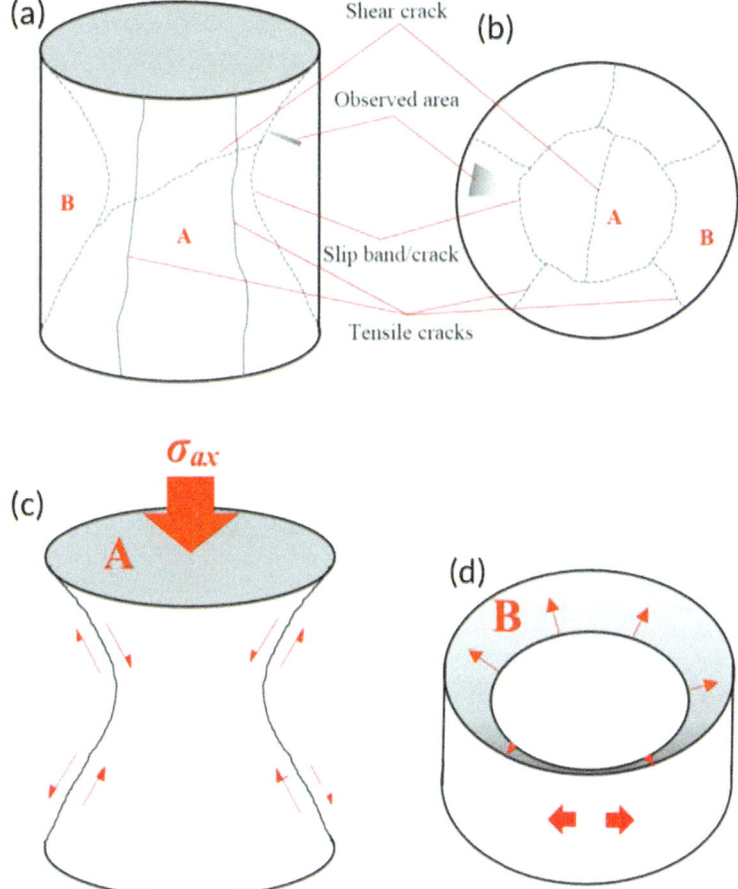

Fig. 4.10 Fractures formed during deformation

with 0~85% stress ratio. A vertical crack was observed in Fig. 4.10b and a crack separating A part and B part was observed in Fig. 4.10a. Judging by the structure, the boundary crack between A and B controls the stability of the sample. SEM is applied to observe the development of boundary cracks. In Fig. 4.11 it is found a crack band and a set of parallel shear cracks. Some of the cracks run along the crystalline boundary, while others, traverse some crystalline grain (Fig. 4.12).

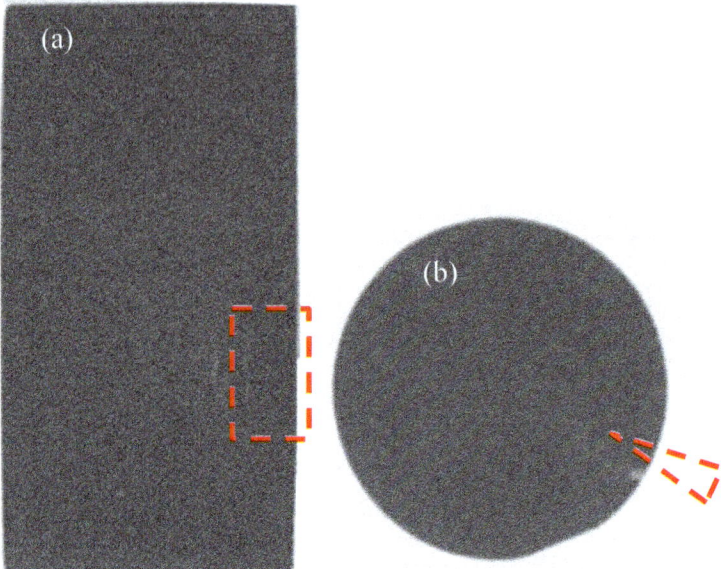

Fig. 4.11 CT images of sample subjected to 3 stress cycles

4.3 Long Interval Effect on Rock Salt

The intervals within 20 min were investigated. It is found that the time interval continuously affects the fatigue properties of rock salt. However, the critical threshold is unclear. It is necessary to continue the tests to find out the longest effective interval.

4.3.1 Experiment Setup

Two cycles were completed in these tests. Firstly, the samples were loaded to 65% of the compression strength with 2 KN/s loading velocity, then unloaded to zero with 5 KN/s loading velocity. The second loading continued to failure with 2 mm/min velocity (Fig. 4.13). The time interval between two cycles Δt took is 1.5 h, 1 h, 2 h, 4 h, 8 h, 4 d. During the interval, samples were wrapped in a sealed plastic bag. The sample and loading machine are the same as previously. These tests are denoted as long discontinuous tests (LDTs). Every test has been performed twice, numbered as *a* and *b*.

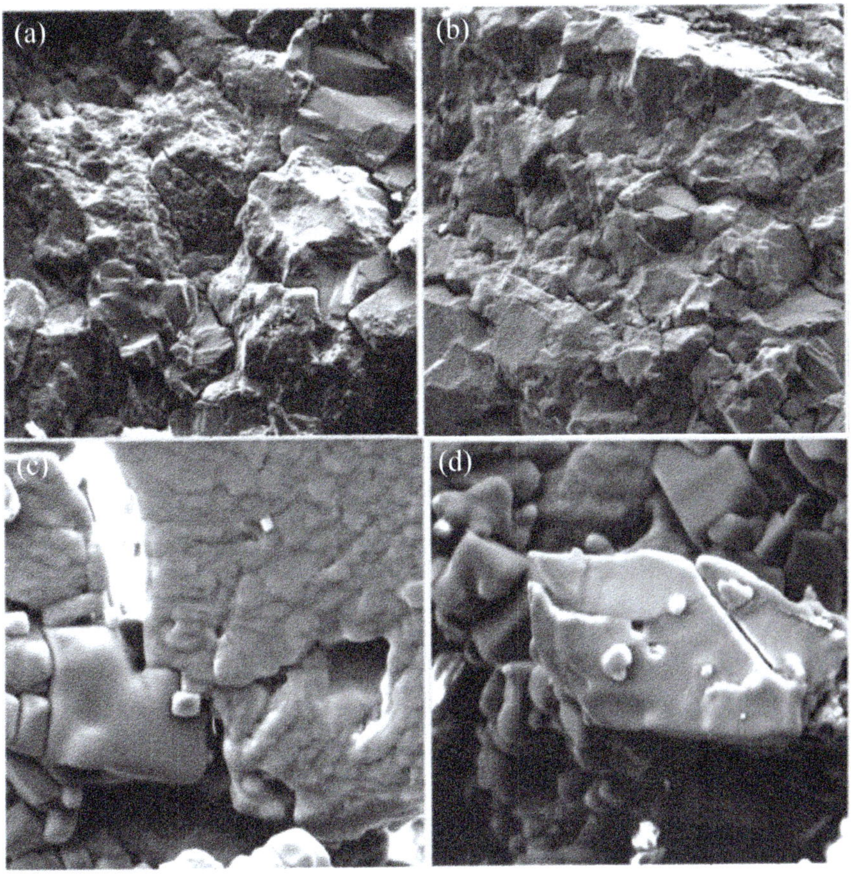

4.3 Long Interval Effect on Rock Salt

Fig. 4.12 SEM images of fracture surface

Fig. 4.13 Loading path of LDTs

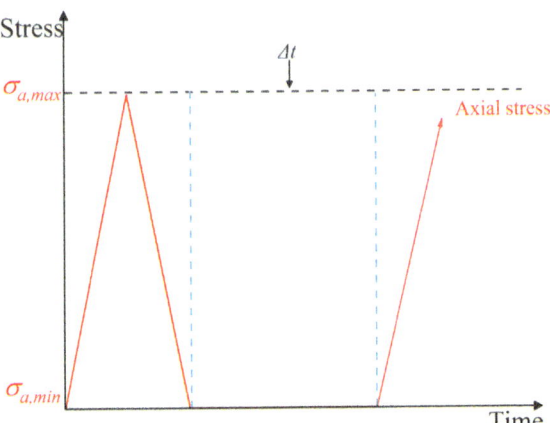

Table 4.1 Peak stress, peak strain and elastic modulus of the sample in second loading

Sample No.	a	b
	Peak stress σ_{pk}/peak strain ε_{pk}/elastic modulus E/Possion's rate μ	
Interval Duration	MPa/100%/GPa/1	
0.5 h	42.46/0.0592/4.202/0.07	41.87/0.0601/3.756/0.05
1 h	40.54/0.0514/4.24/0.07	38.68/0.0483/4.286/0.07
2 h	39.38/0.0575/3.441/0.09	37.97/0.0524/3.912/0.08
4 h	36.97/0.0562/2.563/0.25	37.69/0.0497/3.096/0.21
8 h	38.58/0.0664/2.196/0.17	37.11/0.0504/3.531/0.21
4 d	36.46/0.0527/3.009/0.27	37.62/0.0537/3.400/0.20

4.3.2 Experimental Results in the Long Interval Test of Rock Salt

It is found from the tests that the time interval of 4 h as the threshold takes effect in the second loading. The effect contains:

1. Peak stress of the second loading decreases with the interval (Table 4.1) below 4 h, but fluctuates with the interval above 4 h. Similarly, the elastic modulus (calculated from linear segments in loading process) decreases notably as the interval is less than 4 h, not notably when the interval is larger than 4 h.
2. The distance between unloading curve and the second loading curve becomes wider as the interval becomes longer. That the compaction stage in second loading becomes longer indicates the porosity increases during the intervals. The yield stress of second loading decreases as is shown in Fig. 4.14.

4.4 Discontinuous Fatigue Life Model of Rock Salt

The DFTs shows the effect of time interval on discontinuous fatigue and VCFT shows the laws of the time interval effect. However, both the data quality and quantity of DFTs are not sufficient. Here more discontinuous fatigue tests (DFT2s) with the same as DFTs loading path were conducted to illustrate the relationship between fatigue life and the interval time. The obtained fatigue life is indicated in Table 4.2.

The experimental results show that the fatigue life decreases with interval time. The relationship between the two is shown in Fig. 4.15, a logarithm function.

The possible empirical relationship between fatigue life and interval time is

$$N_f(t) = B_n(\Delta W^p) \lg(\Delta t) + A_n N_f(0) \quad (4.3)$$

$N_f(0)$ is the conventional fatigue life; $N_f(t)$ is the discontinuous fatigue life; A_n is the stress weight; B_n is the interval weight. Since the DFT2 tests are conducted

4.4 Discontinuous Fatigue Life Model of Rock Salt

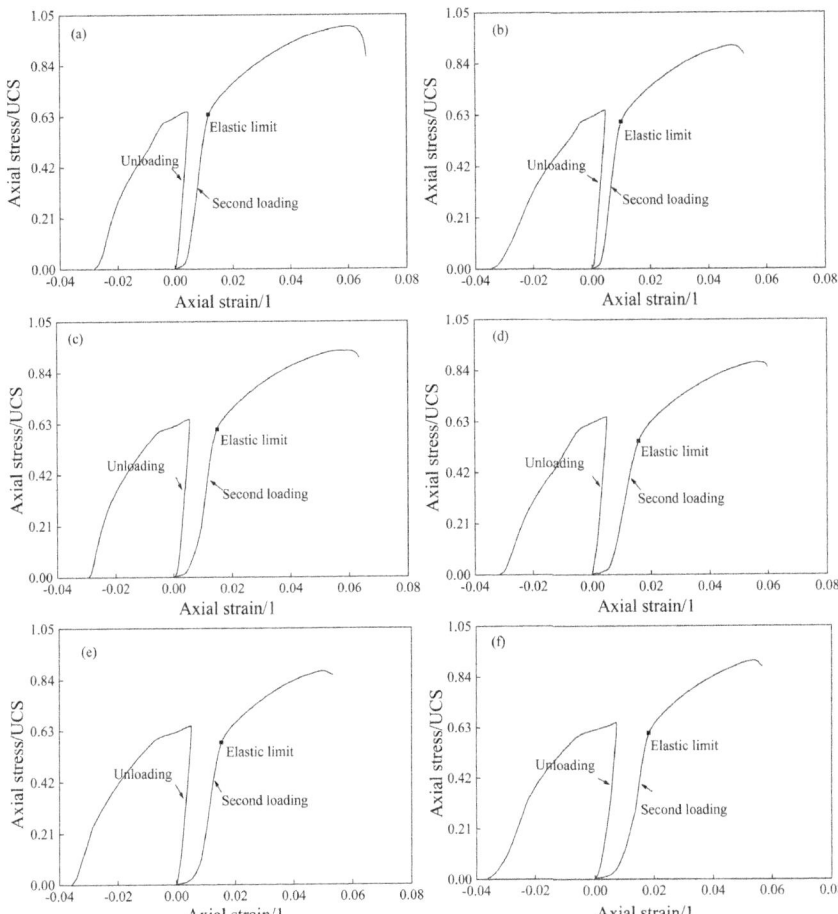

Fig. 4.14 Axial stress-axial strain curve LDTs, **a** $\Delta t = 0.5h$, **b** $\Delta t = 1h$, **c** $\Delta t = 2h$, **d** $\Delta t = 4h$, **e** $\Delta t = 8h$, **f** $\Delta t = 4d$

Table 4.2 Fatigue life of sample in DFT2s

Interval (s)	0	3	5	20	100	300	600	1200	2400
Fatigue life	89	75	52	50	41	34	27	21	18
	88	76	62	48	46	36	33	22	15
	96	70	58	64	47	35	29	26	20
Average	91	74	57	54	45	35	30	23	18
SD (standard deviation)	4.36	3.21	5.03	8.72	3.21	1.00	3.06	2.65	2.52

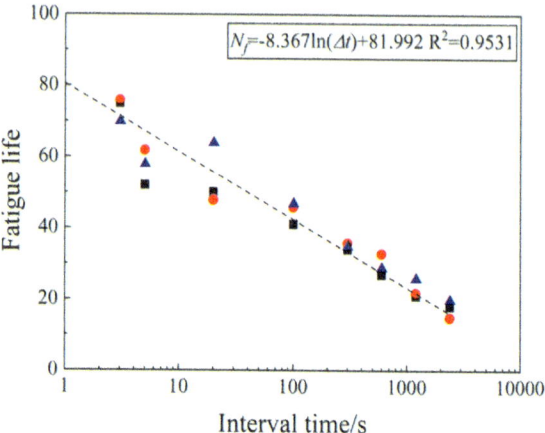

Fig. 4.15 Relationship between interval duration and fatigue life in DFT2s

only with the fix stress limits, A_n and B_n are constant. According to the calculations, it can be obtained that $A_n = 4.58$, $B_n = -34.75$, $\varsigma = 17.61$

$$N_f(t) = B_n \lg(\Delta t) + A_n (R_{max})^{-\varsigma} \tag{4.4}$$

4.5 Conclusions

This chapter investigated the basic features of discontinuous fatigue, threshold of the time interval and the lower limit effect on the discontinuous fatigue. The results of discontinuous fatigue tests shows:

(1) Under the same stress and environmental conditions, the rate of plastic development in discontinuous fatigue is significantly faster than that in conventional fatigue. The fatigue life in discontinuous fatigue is notably shorter than that in conventional fatigue. An increase in the duration of the time interval promotes plastic development in discontinuous fatigue and reduces its fatigue life.
(2) In discontinuous fatigue experiments where the lower stress limit is not zero, the plastic deformation of the rock salt samples develops faster than that of the conventional fatigue samples, and the fatigue life is shorter.
(3) Long-term interval experiments reveal that the critical value of the time interval is approximately 4 h. Within 4 h, the weakening characteristics of rock salt are more evident. After 4 h, the effect of the time interval on the rock salt becomes less significant. The longer the time interval, the lower the yield point of rock salt in subsequent loading, the lower the elastic modulus (loading phase), and the greater the Poisson's ratio (loading phase).

(4) The mechanism of discontinuous fatigue is that the residual stress continues to act during the time interval, causing damage or inducing the Bauschinger effect, which leads to greater deformation and damage in the material during subsequent loading.

References

1. Katarzyna Cyran. Insight into a shape of salt storage caverns[J]. Archives of Mining Sciences, 2020, 65(2).
2. Jinyang Fan, Deyi Jiang, Wei Liu, et al. Discontinuous fatigue of salt rock with low-stress intervals[J]. International Journal of Rock Mechanics and Mining Sciences, 2019, 11577–86.
3. Fan Yang, Zongze Li, Marion Fourmeau, et al. A unified constitutive model for salt rocks under triaxial creep-fatigue loading conditions[J]. Tunnelling and Underground Space Technology, 2024, 154: 106116.
4. Zongze Li, Yanfei Kang, Jinyang Fan, et al. Creep–fatigue mechanical characteristics of salt rocks under triaxial loading: an experimental study[J]. Engineering Geology, 2023, 322: 107175.
5. Xunjian Hu, Kang Bian, Jian Liu, et al. Discrete element simulation study on the influence of microstructure heterogeneity on the creep characteristics of granite[J]. Chinese Journal of Rock Mechanics and Engineering, 2019, 38(10): 2069-2083.

Open Access This chapter is licensed under the terms of the Creative Commons Attribution-NonCommercial-NoDerivatives 4.0 International License (http://creativecommons.org/licenses/by-nc-nd/4.0/), which permits any noncommercial use, sharing, distribution and reproduction in any medium or format, as long as you give appropriate credit to the original author(s) and the source, provide a link to the Creative Commons license and indicate if you modified the licensed material. You do not have permission under this license to share adapted material derived from this chapter or parts of it.

The images or other third party material in this chapter are included in the chapter's Creative Commons license, unless indicated otherwise in a credit line to the material. If material is not included in the chapter's Creative Commons license and your intended use is not permitted by statutory regulation or exceeds the permitted use, you will need to obtain permission directly from the copyright holder.

Chapter 5
Creep–Fatigue Mechanical Characterization of Rock Salt Under Uniaxial Stresses

During the operation of a CAES power plant, there is a spatial and temporal difference in the use of energy between the demand side (households, factories, etc.) and the supply side (wind, solar, etc.). That is, when a sustainable energy source generates electricity and stores it in a salt cavern by means of compressed air, the gas is not immediately released to generate electricity, so there is a certain period of high-pressure stabilization. Similarly, there will be a lower pressure stabilization period for the internal gas pressure after the CAES plant has completed its peaking [1]. The study of simple rock salt fatigue mechanical properties does not fully reflect the real mechanical state of the surrounding rock of the CAES reservoir [2]. On the one hand, the effect of low stress on the fatigue behavior of rocks has been well investigated by scholars around the world. On the other hand, the effect of high stress plateaus, i.e. fatigue behavior under creep–fatigue alternation, has been studied for metals [3–5], asphalt, plastic wood flooring or saturated soft clay [6–9], but is rarely reported on rock materials. Therefore, there is a need to carry out relevant research on the mechanical characteristics of creep–fatigue of rock salt.

5.1 Experimental Methods

The loading scheme for the creep–fatigue test of the rock salt was as follows: first, the upper stress limit is applied and the load was stabilized for one hour defined as T stage before releazed. The purpose of this "first one hour high stress plateau i.e. T-stage" is to simulate the rheological state of the salt cavern surrounding rock due to the long-term high-pressure environment when the salt cavern is first injected with air. Subsequently, a high stress plateau is applied and held for a given duration during every other stress cycle. The purpose of this setup (alternating cycle with and without plateaus) was to allow a comparative analysis of the effects of two different stress cycles, on the same specimen, therefore avoiding the dispersion caused by specimen variability.

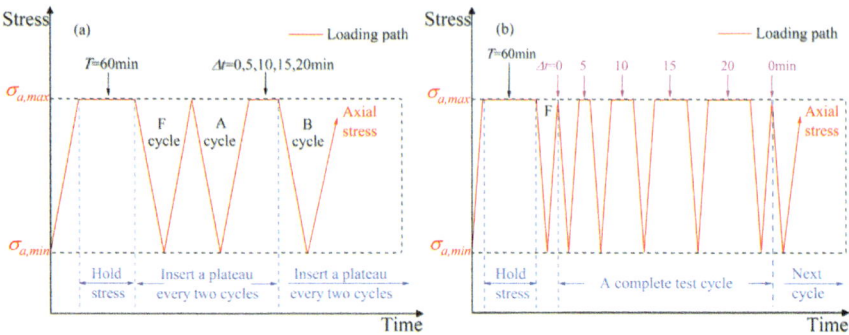

Fig. 5.1 Loading paths for the creep–fatigue tests: **a** CCFs and **b** VCF

Six experiments were carried out. Five conventional creep–fatigue (CCF) tests, and one validation creep–fatigue (VCF) test. The five CCF tests were performed with high stress plateaus lasting 0, 5, 10, 15, and 20 min. Note that the 0 min test is the control group to observe the difference with and without creep. The pre-plateau cycle is defined as the A cycle, and the post-plateau cycle is defined as the B cycle. The first cycle (F cycle) is neither a pre-plateau nor a post-plateau and is called the first cycle. The loading path for CCF tests is illustrated in Fig. 5.1a. During the VCF test, high stress plateaus lasting 0, 5, 10, 15, and 20 min were applied to the same specimen, as shown in Fig. 5.1b. This test was used for validation first, to verify the accuracy of the other five tests and, second, to avoid the influence of testing specimen.

The specific test procedure is as follows. First, place the selected specimen between the two indenters, and close the protection window to ensure the safety of the testing machine. Turn on the power to the test machine and the control computer. Secondly, set the loading program, and apply a certain preload force to the specimen to ensure the fit between the specimen and the indenter. After the preload force is applied, start the program according to Fig. 5.1 procedures. Third, after the specimen is destroyed and the tests stops automatically, remove the specimen and store it in a sealed plastic bag.

The loading and unloading levels and rates of the test were as follows: A computer-controlled program was used to load up to 85% of the uniaxial compressive strength (25 MPa) at a rate of 2 kN/s as the upper limit pressure, held for one hour, and then unloaded down to 3% of the uniaxial compressive strength (1 MPa) at the same rate as the loading rate. Compared to the conditions in a real CAES plant, the upper limit pressure selected for the laboratory test was higher, the lower limit pressure was lower, and the unloading pressure was faster, to be able to complete the test in one working day (12–24 h). Therefore, this experiment is focused on determining the effect of creep on fatigue at the laboratory scale. Each loading path was carried out one time. The main results of the creep–fatigue tests are shown in Table 5.1. The fatigue life is defined as the number of cycles of loading that the specimen undergoes before complete failure, and the creep life is defined as the total time spent under high stress plateau during the test (including the first one-hour high stress period).

Table 5.1 Test results of the CCF tests

Plateau times/min	Fatigue life/cycle	Creep life/min
0	201	60
5	112	275
10	88	430
15	68	495
20	60	580

5.2 Test Results and Analysis

5.2.1 Stress–Strain Curve in the Uniaxial Creep–Fatigue Tests

The fatigue stress–strain curves for the 0, 5 and 10 min CCF tests and VCF test are shown in Fig. 5.2. The common feature of the four curves is that the rock salt undergoes a large deformation after the first loading. After the one-hour first high stress plateau, for the CCF test at 5 min, as an example, the specimen enters the creep–fatigue cyclic loading stage, and the fatigue curve shows a comb-dense-comb development pattern on the whole. This is similar to conventional fatigue [10], indicating that the high stress plateau does not change the essential characteristics of the fatigue curve. However, its first stage (deceleration deformation stage, **I**) is indeed greatly shortened. Previous studies [11] have found that we can roughly consider the first 20% of cycles of fatigue damage of rock salt as the deceleration deformation stage, but in this test, the deceleration deformation stage accounts for only approximately 10% of the whole cycle. At the same time, the distinction between the deceleration deformation and the stabilization deformation (the second stage, **II**) stages of the specimen is not obvious because after the loading stage, the specimen undergoes another hour of high stress, the internal cracks of the specimen are compacted and compressed, and the integrity of the specimen is improved. When entering the third stage (accelerated deformation stage, **III**) near the damage, the curve becomes sparse again, and not only the deformation in the high stress plateau but also the residual strain generated by the cyclic loading increases (the law of residual strain will be described in Sect. 2.3.22.3.2). The specimens from the 5 min CCFT eventually broke down during the last high stress plateau. A similar pattern of development was observed for the remaining five sets of tests; the 0 min CCF test (i.e., conventional fatigue) specimens were damaged in the loading stage of the last cyclic load, and the damage stress was slightly less than the upper limit stress, which is consistent with the majority of the test phenomena, indicating that even though the test was designed for an initial one-hour high stress plateau, it did not change the nature of the fatigue test.

For the four CCF tests, the total test time is different for two reasons: the number of cycles before failure is different (larger for shorter plateau duration), and the plateau duration varies between tests (5–20 min). Therefore, it is not really straight forward

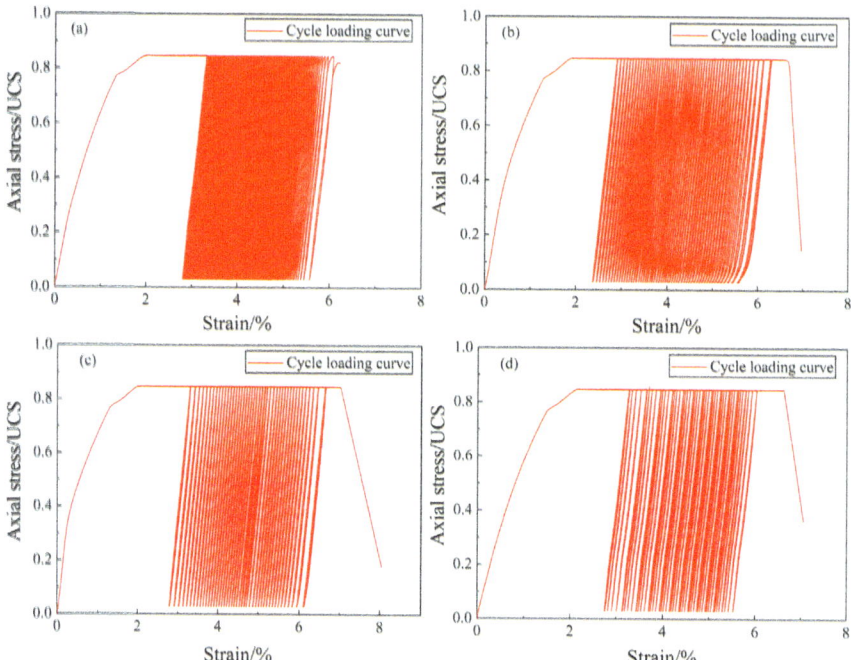

Fig. 5.2 The stress–strain curves of the **a** 0 min plateau time CCF test, **b** 5 min plateau time CCF test, **c** 10 min plateau time CCF test, and **d** is the VCF test

to interpret the experimental results in term of total duration. Instead, taking the time when the 5 min CCF test enters the third stage as the calculation limit, calculate the difference between the strain of the specimens in the four CCF tests at the same test time and the strain when they first entered the creep–fatigue alternating stage, as shown in Fig. 5.3.

From the Fig. 5.3, it can be observed that under the same test duration of 5, 10, 15, and 20 min for the CCF tests, the corresponding strain differences of the rock salt specimens exhibit a decreasing trend and show a nearly linear relationship. This phenomenon can be attributed to the fact that under the same test duration, shorter high-stress plateau durations result in more cyclic loads applied to the specimens. In the early stage of the tests, the cracks generated by cyclic loading are compacted during the high-stress interval. As the tests progress, these cracks gradually evolve from small cracks to large ones, and the high-stress interval transforms from promoting crack closure to driving crack propagation and sliding of fracture surfaces. This leads to an accelerated deformation rate of the specimens, and they quickly enter the third stage, ultimately leading to failure.

5.2 Test Results and Analysis

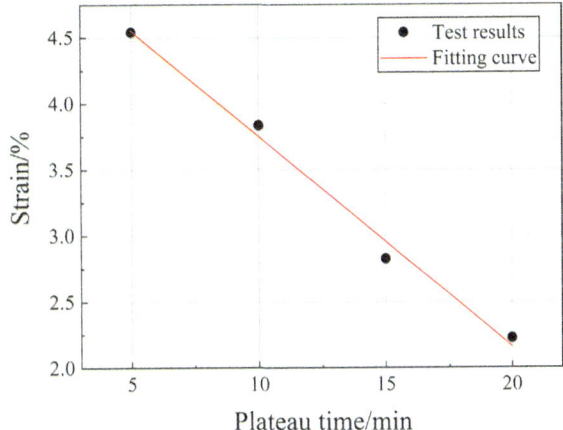

Fig. 5.3 The strain difference at the same test time for rock salt in the 5, 10, 15 and 20 min CCF test

5.2.2 Creep–Fatigue Strain Rate in Rock Salt

In addition, the deformation rate for the high stress creep period for each test were calculated to observe the pattern of variation in creep rate (v_c) at a given plateau time with the following equation:

$$v_{ci} = \frac{\varepsilon_{ci}}{t_i} \tag{5.1}$$

where i is the number of high-stress cycles excluding the initial one hour high stress period (for this one hour high-stress period, v_c is calculated by dividing it into 4 equal parts), ε_{ci} is the creep strain difference at the high stress plateau and t_i is the plateau time (for the initial one hour high stress period, the high stress period is divided into four time points bounded by 200, 1000, 1800, 2600, and 3400). For the high stress plateau during creep–fatigue, v_c is calculated by removing 10% of the creep time before and after. The purpose is to eliminate errors in the alternating periods of acceleration/unloading and creep. The v_c values for the 5 min CCF test are shown in Fig. 5.4.

The figure shows that the v_c of the rock salt specimens shows a decreasing trend during the first hour-long high stress plateau, but the decreasing trend slows, which is consistent with the known results of rock creep tests. This indicates that a rock salt specimen is in the first stage of creep, i.e., deceleration creep, reflecting the rapid closure of internal microfractures/pores, when subjected to external loading, leading to an increased integrity of the specimen. After the first cyclic loading (i.e., fatigue action), we observed an increase in v_c during the first alternating creep–fatigue action (as shown in the blue box of Fig. 5.4). This indicates that fatigue has a positive effect on creep because after the withdrawal of pressure and reloading, some of the fractures and structural surfaces inside the rock specimen that had been stabilized were opened again, resulting in a jump in v_c (i.e., the rock salt specimen is considered to have

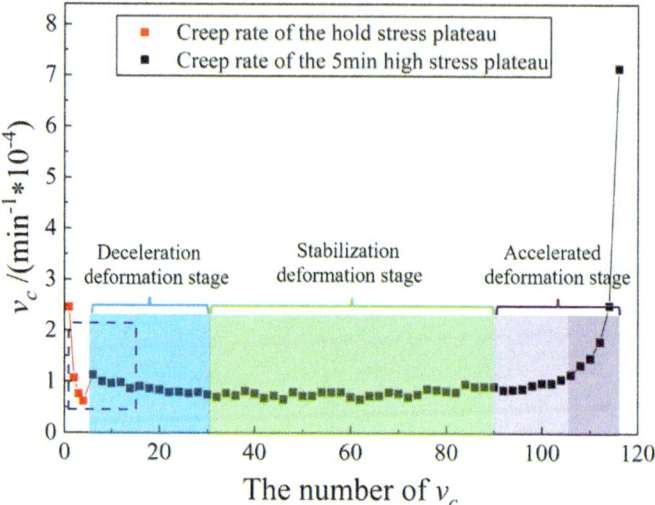

Fig. 5.4 Creep rate during the 5 min plateau time CCFT

restarted the creep process with a larger creep rate, similar to the process in the red part). Describe what is in blue, green and purple stages before the last 20%. Entering the last 20% of the cycle, the v_c of the rock salt increases again, and the rock salt specimen enters the third stage of creep, commonly called accelerated creep stage. At this time, the microfractures inside the rock salt develop rapidly, and the fracture surface gradually penetrates to form a macrofracture surface. During the last 5% of the cycle, v_c shows an exponentially rising trend. Finally, in the last creep period, the rock salt specimen is structurally destabilized and damaged. The v_c exceeded 7 × 10^{-4} min^{-1}, far exceeding the initial v_c of 2.5 × 10^{-4} min^{-1}.

To investigate the effect of fatigue on creep for different high stress plateau times (5–20 min), we calculated the average creep rate ($\bar{v}_c$) and average creep strain ($\bar{\varepsilon}_c$) during the second stage by using Eqs. (5.2) and (5.3):

$$\bar{v}_c = \frac{\sum_n^m v_{ci}}{m-n} \quad (5.2)$$

$$\bar{\varepsilon}_c = \frac{\sum_n^m \varepsilon_{ci}}{m-n} \quad (5.3)$$

where n and m are the beginning and end loading cycle numbers of the second stage in every CCF test. The results are shown in Fig. 5.5.

We found that as the duration of the high stress plateau increases, the average creep rate decreases, indicating that the effect of fatigue on creep gradually weakens. However, there is a threshold value for this phenomenon, i.e., the creep rate for 5, 10, and 15 min shows a linear decrease, but the creep rate for 15 and 20 min, decreasesless,

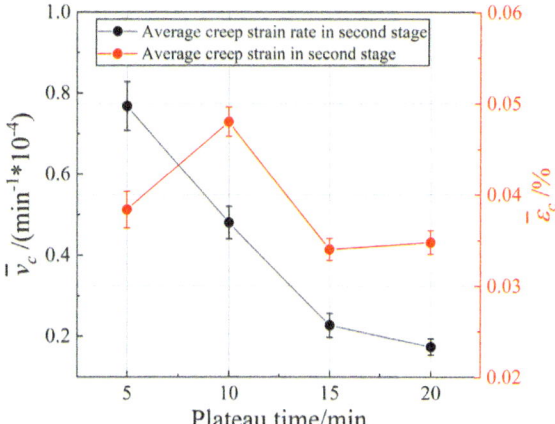

Fig. 5.5 Average creep rate and strain during the stabilization deformation (the second stage, **II**) under different plateau times

indicating that the effect of fatigue on creep stabilizes when the plateau duration is longer than 15 min. It increases during 5 and 10 min plateau, decreases between 10 and 15 min, and increases slightly again between 15 and 20 min CCF test. In a first analysis, this would suggest that the 15 min plateau time would be an optimum for rock salt samples in the CCF test. However, for guiding recommendations regarding salt cavern storage, it is necessary to provide them based on tests conducted at the engineering scale, so we prefer not to draw any strong conclusions here. The creep deformation and strain rate for the VCF test was calculated with the same method, showing the same trend as in the CCF test, i.e. the average creep rate and average creep strain at each cycle (0–20 min duration) correspond to the values obtain for the corresponding CCF test.

5.2.3 Creep–Fatigue Residual Strain in Rock Salt

Rocks are deformed when subjected to external loads, and the deformation includes reversible elastic deformation and irreversible plastic deformation (i.e., residual strain). When a rock is subjected to cyclic loading, plastic deformation will occur with each cycle. In a creep–fatigue test, creep deformation and fatigue deformation together lead to plastic deformation of the rock salt, and when the plastic deformation accumulates to a critical state, the rock will become damaged.

The residual strain (ε_r) per cycle for the creep–fatigue test is calculated by the following equation:

$$\varepsilon_{ri} = \varepsilon_{ai} - \varepsilon_{bi} \tag{5.4}$$

where i is the number of cycles (excluding the first cycle), ε_{ai} is the axial strain at the start of the current cycle (start of the unloading of a cycle), and ε_{bi} is the axial strain at the end of the same cycle (end of the loading of the cycle).

The residual strain ε_r patterns for the control group (i.e., the specimens tested during the 0 min) and 5 min CCF tests, are shown in Fig. 5.6. We found that, compared with conventional fatigue, the ε_r of the control and high stress plateau groups do not have obvious boundaries between the first and second stages and enter the second stage faster, which is similar to the law of fatigue curve development, for reasons analysed in the previous section (not be repeated here). Another obvious difference is that the ε_r of A and B cycles of the control test (no stress plateau) are not significantly different, while the ε_r of A and B cycles of the specimens undergoing fatigue combined with high stress plateaus are significantly different, with ε_r of A cycles significantly larger than those of B cycles. This difference was observed throughout the two first stages of the test process, while this difference is less clear during the third stage of the test. This pattern was found for the other three fatigue test groups with 10, 15, and 20 min plateaus.

Considering that A cycle and B cycle of the creep–fatigue test are always in different positions (the A cycle is always in the even position, and the B cycle is always in the odd position, including the first cycle), the ε_r of the A and B cycles at the corresponding positions were solved by using the arithmetic mean method, and the calculation equation s are consistent with Eqs. (4.1) and (4.2) in Chap. 4 and will not be repeated here.

Four different plateaus were set up for the tests, and each set of tests showed that ε_r of A cycle were greater than the ε_r of B cycle. However, the effect of different plateaus on the development of ε_r varies. We calculated the average value of the difference between the ε_r of A cycle and B cycle during the second stage (stable deformation stage) for the various plateau durations, using Eq. (5.5):

$$\Delta \bar{\varepsilon}_r = \frac{\sum_n^m \left(\varepsilon_{r,An} - \varepsilon_{r,Bn}\right)}{m - n} \tag{5.5}$$

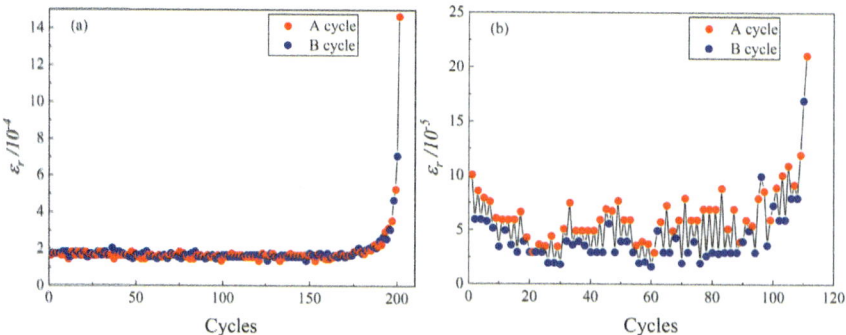

Fig. 5.6 Calculated axial residual strain from the (a) 0 min plateau time CCF test and (b) 5 min plateau time CCF test

5.2 Test Results and Analysis

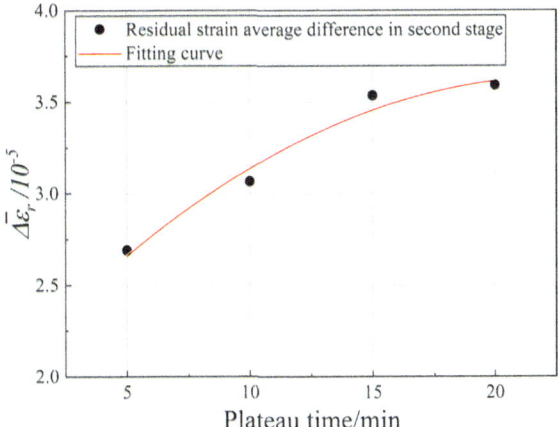

Fig. 5.7 Average values of the residual strain difference between A and B cycles under different plateau times

where $\Delta\bar{\varepsilon}_r$ is the average value of the difference between the ε_r values of A cycle and B cycle from Eqs. (4.1) and (4.2), and n and m are the beginning and end loading cycle numbers of the second stage. The calculation results are shown in Fig. 5.7.

We found that the average value of the residual strain difference increases gradually with increasing high stress plateau time. This shows that the effect of creep on fatigue increases with increasing plateau time. Additionally, there seem to be a threshold in this increase, consistent with the conclusions obtained in Sect. 2.3.2 indicating that the effect of creep on fatigue is also stabilizing. The interaction between creep and fatigue is a positive interaction. We discuss the specific reasons for this pattern below.

5.2.4 Relation Between Fatigue Life and Creep Life in the Creep–Fatigue Test of Rock Salt

In the process of rock salt creep–fatigue testing, the damage sustained by rock salt is constituted by both fatigue damage and creep damage. Regarding fatigue damage, the initiation and propagation of rock fatigue cracks are influenced by its micro-fissure forms. Rock fatigue failure fractures include transgranular fractures, intergranular fractures, and a combination of transgranular and intergranular fractures. Different mineral crystals within the rock exhibit distinct fracture patterns [12]. For instance, in granite, where feldspar, quartz, and mica crystals coexist, transgranular fractures primarily occur in feldspar crystals under fatigue loading. Quartz crystals experience a limited amount of both transgranular and intergranular fractures, while mica crystals mainly undergo intergranular fractures. The loading pattern for fatigue involves cyclic loading and unloading, where the dynamic effects and energy changes resulting from loading and unloading promote transgranular fractures. Moreover, due to the presence of grain boundary defects, cracks can nucleate at weak points along the grain boundaries, leading to intergranular fractures. When transgranular crack initiation is close to intergranular crack nucleation, coupled transgranular-intergranular fractures may occur. Among these three types of micro-cracks, intergranular fractures exhibit the fastest growth rate, followed by transgranular fractures [13]. Intergranular cracks tend to develop in the direction of loading, whereas the growth direction of transgranular cracks is more random. During reverse unloading, stress concentration occurs at the crack tip area, inducing significant residual tensile stress and promoting crack extension [14]. Scanning electron microscope observations of failure surfaces reveal that the primary micro-mechanism for rock salt fatigue testing is cleavage failure, with shear failure as a secondary factor. Intergranular and transgranular cracks coexist within the fracture surface [15]. Under cyclic loading and unloading, the microstructure of the rock repeatedly undergoes deformation, with more movement occurring along grain boundaries, resulting in intergranular fractures. The cohesion between particles at grain boundaries consequently decreases, leading to particle detachment. Detached particles intensify this frictional movement, causing further particle detachment and generating visible macroscopic debris [16]. The initiation and development of fatigue cracks are also influenced by the presence of primary cracks, which, under stress concentration, serve as starting points for rock failure crack formation [17]. With an increase in the number of fatigue cycles, more tip cracks are formed, and internal cracks transition from straight to curved as deformation progresses [18].

The mechanism of creep damage is somewhat different. The process of rock creep deformation failure involves crack initiation, growth, bifurcation, and coalescence, ultimately leading to rock rupture [19]. Cracks originate from the formation of creep micropores and the diffusion of micro-defects [20]. In the vicinity of rock grain boundaries, stress concentration often occurs due to the discordant movement of grains and the presence of impurities, leading to pore formation. Pores continue to grow under constant loading, eventually connecting with surrounding pores to form intergranular micro-cracks. These micro-cracks then further connect and converge to form main cracks. Secondary cracks develop around the main crack, and creep fractures mainly develop along the direction of the principal stress (or a smaller angle) [21]. In the case of crystalline minerals like rock salt, there are internal defects such as grain vacancies [22], dislocations, and grain boundaries, which, under constant loads, temperature, and humidity, undergo diffusion and subsequently initiate cracks [23]. The entire process of creep crack initiation, development, and coalescence is influenced by the microstructural heterogeneity of the rock. The unevenness of the rock causes differences in the location of crack initiation and the rate of crack propagation and widening. The presence of impurity particles within the rock, the variation in grain size, and the orientation of grains all directly affect crack initiation. Different deformation moduli and Poisson's ratios of various grains and impurities result in the need for the microstructure of the rock to adjust its motion over time under load, and the initiation and development of cracks are outcomes of the mutual adjustment and coordination of the rock's microstructural motion [24].

Few experimental studies on creep–fatigue of rocks have been conducted, but research on creep fatigue of metals has been ongoing for nearly 60 years [25–28]. Here, drawing on the statistical method of metal creep–fatigue life, the fatigue life

5.2 Test Results and Analysis

is defined as the number of cycles of loading that the specimen undergoes before complete failure, and the creep life is defined as the total time spent under high stress plateau during the test (including the first one-hour high stress period). the fatigue life for various plateau time is shown in Fig. 5.8.

We found that with the increase in plateau time, the fatigue life decreases and satisfies a linear equation:

$$F_l = 126 - 3.52P_t, \ R^2 = 0.93 \tag{5.6}$$

The creep life for various plateau time is shown in Fig. 5.9.

With the increase in plateau time, creep life increases and satisfies the following linear equation:

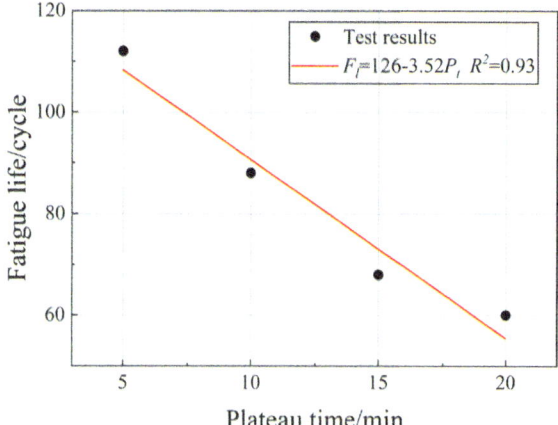

Fig. 5.8 Fatigue life for different plateau times

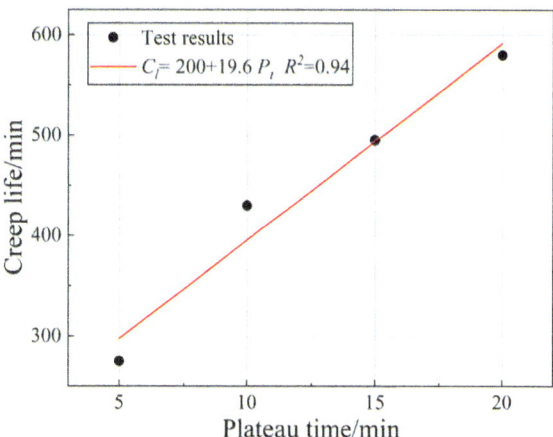

Fig. 5.9 Creep life for different plateau times

$$C_l = 200 + 19.6P_t, \quad R^2 = 0.94 \tag{5.7}$$

This is close to the conclusion reached by the creep–fatigue testing of metals. According to a study by Joseph Oldham et al. [29], the damage equation of metallic materials during creep–fatigue damage can be defined as follows:

$$\text{Fatigue damage}: \sum_\alpha \left(\frac{N}{N_d}\right) \tag{5.8}$$

$$\text{Creep damage}: \sum_\beta \left(\frac{T}{T_c}\right) \tag{5.9}$$

For materials damaged during creep–fatigue experiments, fatigue and creep damage are cumulative processes. If fatigue and creep separately affect the material during the test, the damage equation can be written as:

$$\sum_\alpha \left(\frac{N}{N_d}\right) + \sum_\beta \left(\frac{T}{T_c}\right) = 1 \tag{5.10}$$

However, in the American Society of Mechanical Engineers (ASME) code [30], there is a positive interaction assumed between creep and fatigue in metallic materials during the creep–fatigue test. Therefore, Eq. (5.10) is modified as follows:

$$\sum_\alpha \left(\frac{N}{N_d}\right) + \sum_\beta \left(\frac{T}{T_c}\right) < 1 \tag{5.11}$$

Further analysis revealed that a correlation term could be used to represent the interaction of creep and fatigue.

$$\sum_\alpha \left(\frac{N}{N_d}\right) + \gamma \left[\mu\left(\frac{N}{N_d}\right) + \theta\left(\frac{T}{T_c}\right)\right]^\delta \sum_\beta \left(\frac{T}{T_c}\right) = 1 \tag{5.12}$$

where N is the number of cycles, N_d is the total number of cycles in the test under the α type, T is the creep time, T_c is the total creep time in the test under the β type, and γ, μ, θ, and δ are the coefficients related to creep–fatigue interactions.

For a given test, the test conditions and environment were determined. Therefore, some scholars [7, 31] have simplified Eq. (5.12) by simplifying the creep–fatigue correlation term to $B\left(\frac{N}{N_d} \times \frac{T}{T_c}\right)^{\frac{1}{2}}$. A new equation is proposed:

$$\frac{N}{N_d} + B\left(\frac{N}{N_d} \times \frac{T}{T_c}\right)^{\frac{1}{2}} + \frac{T}{T_c} = 1 \tag{5.13}$$

5.3 Mechanisms of Creep–Fatigue Interactions in Rock Salts

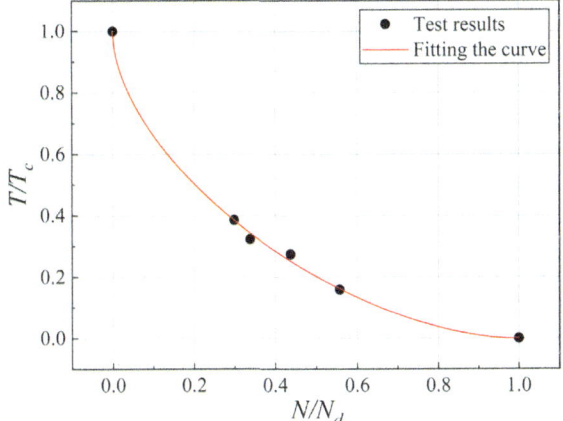

Fig. 5.10 Normalized creep life and fatigue life for creep–fatigue tests (5, 10, 15, 20 min CCF test) on rock salts, with fitting curve

where B is the correlation coefficient, and a larger value indicates that the creep–fatigue interaction is stronger. Figure 5.10 show the normalized fatigue life and creep life for the 4 different tests (5, 10, 15, 20 min CCF test). For rock salt, a B value of 1.01 is obtained. This indicates the existence of creep–fatigue interactions, which is consistent with the experimental phenomenon. There are many uncertain structures within our rock, as a natural material, that are difficult to observe and resulting in a relatively dispersed mechanical property. Moreover, the tests reflect that the interactions between creep and fatigue are not consistent. A more accurate equation is given as follows:

$$\frac{N}{N_d} + A\left(\frac{N}{N_d}\right)^\varphi \left(\frac{T}{T_c}\right)^\omega + \frac{T}{T_c} = 1 \quad (5.14)$$

where A is the interaction coefficient and φ and ω are the fatigue and creep damage indices, respectively, which reflect the degree of fatigue damage and creep damage on the creep–fatigue interaction damage. We speculated that for this experiment, A can be adopted as the hardening factor of creep on the fatigue residual strain, and B can be adopted as the catalytic factor of fatigue on creep rate. We will complete more tests to finalize the values of these coefficients.

5.3 Mechanisms of Creep–Fatigue Interactions in Rock Salts

In pure fatigue testing of rock salt, there is no significant difference in the development of residual strain per cycle except for the traditional three stages [32]. In the low-stress-interval fatigue test, the residual strain in the post-interval cycle is greater than the residual strain before the interval [33]. The residual stresses during the low stress interval were considered to be the cause of this phenomenon. In the creep–fatigue experiments completed by the authors, however, the opposite phenomenon occurred, with the residual strain before the high stress plateau being greater than the residual strain after the plateau. The fatigue cycles can also affect the creep rate and strain of the rock salt. We think that the following reasons are responsible for this phenomenon.

Factors affecting the deformation mechanism of the material include the internal structure and the external environment. The internal structure includes the grain size and impurity distribution, and the external environment includes the temperature, stress state, and hydrogeological environment. It has been pointed out that the creep of rock salts can be divided into dislocation creep (Fig. 5.11b) and pressure solution creep (Fig. 5.11c) [34, 35]. Dislocation creep mainly occurs inside the rock salt crystals, while pressure dissolution creep mainly occurs at the grain boundaries of the rock salt.

The main influence of pressure dissolution creep, in addition to the fundamental factor of stress, is mainly related to the moisture conditions between the rock salt grain boundaries and to the temperature. In the presence of moisture at the rock salt grain boundaries, the rock salt crystals in a high-stress region dissolve, are transported through the fluid, and finally deposit in a low-stress region [36]. Pressure dissolution creep [37–39] is accompanied by the relative dissolution, sliding and rotation of the grains, which gradually eliminates the pores. The temperature level also determines the rate of pressure dissolution creep.

When the temperature is very high, the stress is very low when the creep rate is proportional to the stress, this creep and dislocation relationship is not large, the deformation is mainly caused by the directional flow of the material under the action of stress, this creep is called diffusion creep.

If there are no atoms at a certain equilibrium position in the atomic arrangement, such as the hollow circle shown in Fig. 5.12a, the place is called a hole. The atom on the left, under the action of thermal perturbation, crosses the energy barrier state shown in Fig. 5.12b and results in the configuration shown in Fig. 5.12c, where the atoms move one equilibrium position to the right. When a large number of atoms make this movement, it becomes diffusion. When diffusion is directional, it causes macroscopic deformation.

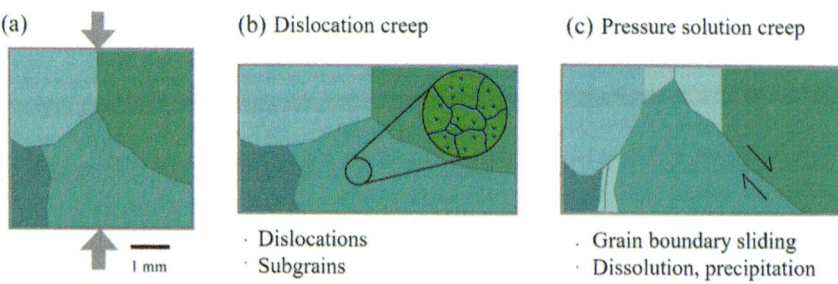

Fig. 5.11 Schematic diagram of rock salt dislocation creep and pressure dissolution creep

5.3 Mechanisms of Creep–Fatigue Interactions in Rock Salts

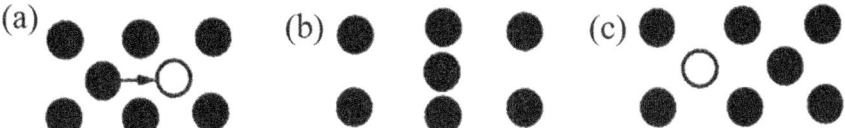

Fig. 5.12 Diffusion caused by hole

In polycrystalline materials, according to the preferential path of atoms, diffusion can be categorized into grain boundary diffusion or bulk diffusion. Grain boundary diffusion refers to the diffusion of atoms moving along the grain boundaries, grain boundary diffusion dominated creep is called Coble creep. Bulk diffusion is the movement within the grain, also known as Nabarro-Herring creep. Under the action of external forces, diffusion is directional. Since at relatively low temperatures, the energy barriers are generally smaller at grain boundaries than within the lattice, grain boundary diffusion is the main causative agent. Whereas at higher temperatures, more atoms within the grain bulk diffuse and its induced strain dominates. The strain rate induced by grain boundary diffusion can be modelled as:

$$\frac{d\varepsilon}{dt} = k_1 \frac{\sigma V}{KT} \frac{D_{gb}\delta}{d^2} \tag{5.15}$$

where, k_1 is the coefficient related to the shape of the grain boundary, V is the volume occupied by a single atom, D_{gb} is the grain boundary diffusion coefficient, δ is the thickness of the diffusion layer at the grain boundary, and d is the average diameter of the grain. The strain rate due to bulk diffusion can be modeled as:

$$\frac{d\varepsilon}{dt} = k_2 \frac{\sigma V}{KT} \frac{D_v}{d^2} \tag{5.16}$$

where, k_2 is a coefficient related to grain shape and D_v is the bulk diffusion coefficient.

From Eqs. (5.15) and (5.16), it can be seen that the strain rate of creep caused by diffusion is linearly related to the stress. It can be inferred that the main mechanism of viscous creep is diffusion. Therefore, viscous creep is also often called diffusion creep. On the other hand, it can be seen from Eqs. (5.15) and (5.16) that if the grain is refined, i.e., if d is reduced, the increase in creep rate due to grain-boundary diffusion is greater than that due to body diffusion, and thus will dominate.

Heard et al. [40]. found experimentally that the conditions under which diffusion creep occurs in rock salt are: (1) the stress is lower than 0.45 MPa, which makes the dislocation density in the crystal very low, and the dislocation motion contributes very little to the total deformation; (2) the temperature is sufficiently high to make the atomic diffusion rate very fast, and the deformation is mainly generated by the directional flow of the atoms.

Since the factors affecting the pressure dissolution creep are well controlled (the test rock salt specimens were dried at a low temperature of 60 °C for 24 h before the test, and the temperature and humidity of the test environment were kept relatively constant during the experiment), the pressure dissolution creep has less influence on the test variables. However, we still observed pressure dissolution creep at some of the grain boundaries, as shown in Fig. 5.13.

Dislocation creep [41, 42] is a more complex form of creep associated with innate crystalline defects within the rock salt, including point defects, line defects, and surface defects.

Here is a brief description of point defects as an example. Point defects are lattice defects of atomic scale magnitude, also known as zero-dimensional defects. During the crystal formation process or under certain physico-mechanical conditions, such as high-energy radiation or thermal activation, point defects are easily formed within the crystal. As shown in Fig. 5.14 there are three types of point defects: Schottky defects (denoted as S defects), Frenkle defects (denoted as F defects), and heterogeneous replacement atoms.

In previous studies [43], researchers have found large discrepancies between the theoretical strengths of crystalline materials and their experimentally measured values, a contradiction that led to the discovery of dislocations in crystals and the development of related theories. When a crystal is subjected to the crystallizing effects of impurities, temperature changes, or vibrational stresses, the arrangement of its internal plasmas is distorted, and the atomic arrangements slide against each other, no longer conforming to the regular arrangement of an ideal crystal, leading to defects called dislocations. More specifically, a dislocation is the demarcation point between the slipped and unslipped areas. The surface on which the relative slip occurs is called the slip surface, the direction in which the slip occurs is called

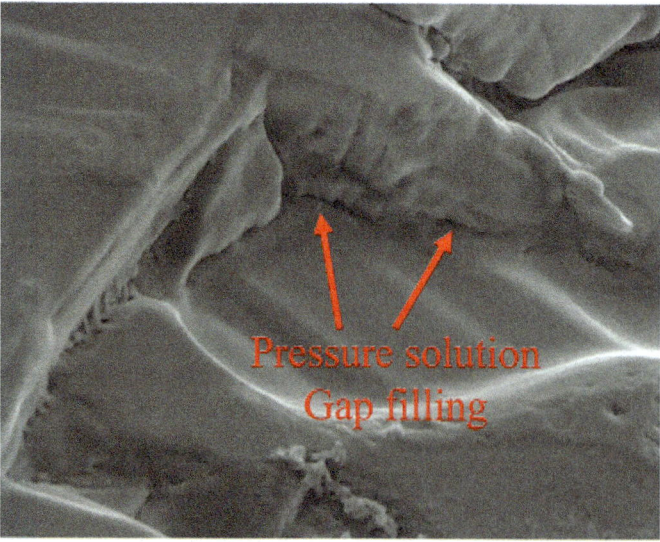

Fig. 5.13 Rock salt pressure dissolution boundary traces (as shown by the red arrow)

5.3 Mechanisms of Creep–Fatigue Interactions in Rock Salts

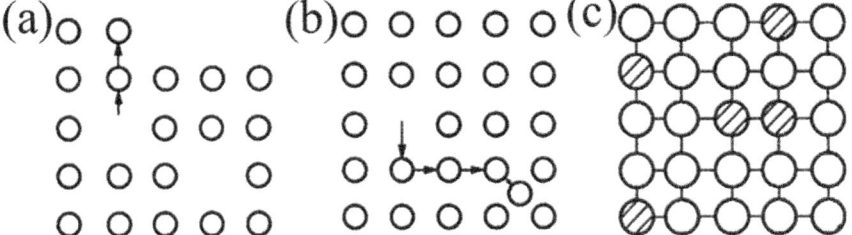

Fig. 5.14 S defect **a**, F defect **b** and heterogeneous replacement atom (**c**)

the slip direction, and the boundary between the slipped and unslipped regions is called the slip line. The simplest dislocations can be categorized into two types: edge dislocations and helical dislocations, and the difference is linked to the direction of the dislocation line and the direction of the dislocation movement under the action of external forces. As shown in Fig. 5.15a, assuming that the crystal undergoes a relative slip in the interatomic distance in the upper and lower parts of the region ABCD, the dividing line between the slipped and unslipped regions is AD, which is the so-called dislocation line that forms an irregular arrangement of atoms. When the direction of slip is perpendicular to the dislocation line called edge-type dislocation; when the direction of slip is parallel to the dislocation line, such as Fig. 5.15b, then the screw dislocation. If the boundary of the slip region is a curve AC, as in Fig. 5.15c, then in different parts of the curve, there are screw dislocations parallel to the curve, perpendicular to the curve edge-type dislocations as well as a mixture of both parallel and perpendicular components of the motion, called mixed dislocations.

Motion is an important aspect of the nature of dislocations; without motion of dislocations, there would not be plastic deformation of crystals, and the ease of dislocation motion is directly related to the strength of crystals. In rock salt, as a typical homogeneous crystalline material, dislocations are commonly used to explain its creep deformation. For inelastic deformation, dislocations are the most important crystal defects in rock salts [44].

When subjected to external loading, the dislocations within the crystal will slip or climb to achieve creep deformation [45]. During this process, the crystals of rock salts will gradually subcrystallize, and the diameter of the subcrystals shows a negative correlation with the deviator stresses applied to the crystals. The crystal

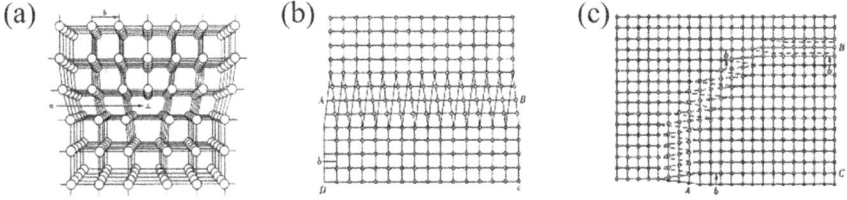

Fig. 5.15 Edge dislocation (**a**), screw dislocation (**b**) and fixed dislocation (**c**)

changes are shown in Fig. 5.16. Adjacent grains may have different dislocations, and their intersections are called grain boundaries. Smaller crystals with different dislocations within the grains are subcrystals, and the interface between the grains is called the subcrystal boundary. The deformation of rock salt crystals includes the deformation of the grains themselves, as well as the deformation caused by the slip of the grain boundaries. Neighboring grains may have different dislocations and their intersections are called grain boundaries. A grain boundary is a natural facet defect. Saltstone is a polycrystalline composite medium consisting of many grains, each of which is a small single crystal. Neighboring grains have different orientations and their intersections are called grain boundaries. The interior of the crystal grains often consists of many crystal blocks with small differences in orientation, and the interfaces between the blocks are called subgranular boundaries. The structure of the crystal surface is different from that of the crystal interior. Since the surface is the termination surface of atomic arrangement, there is no bonding of atoms in the solid on one side, and its coordination number is less than that of the crystal interior, which leads to the deviation of surface atoms from their normal positions and affects the neighboring layers of atoms, resulting in a point distortion that is higher than that in the crystal interior in terms of energy. The deformation of rock salt crystals includes the deformation of the grains themselves, as well as the deformation caused by grain boundary slip.

The results of related tests show that the greater the difference in orientation between two crystals, the greater the amount of slip; in this case, the creep rate increases. While primary slip occurs at the grain boundary, secondary slip may also occur in the grain boundary affected zone (GBAZ) [46] due to stress concentration, and both temperature and stress can affect the size of GBAZ [47]. At the same time,

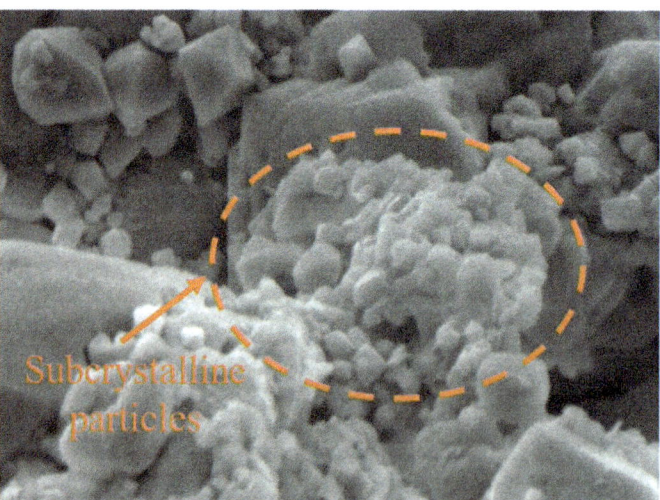

Fig. 5.16 Large number of subcrystalline particles produced by dislocation creep (orange dashed oval)

5.3 Mechanisms of Creep–Fatigue Interactions in Rock Salts

if the difference in orientation between the two crystalline surfaces is smaller, the amount of slip is smaller, which means that the creep rate decreases. During the creep–fatigue experiments [48], the unloading and reloading of stress changes the inherent arrangement between the rock salt crystals in the high stress plateau, leading to crystal disorder (Fig. 5.17). As the difference in orientation between the crystals increases [49], the rate of rock salt creep increases in our analysis of the fatigue-accelerated creep rate. At the same time, as dislocation creep proceeds, dislocation interception and plugging will prevent the dislocation from continuing to slip [50]; to make the dislocation move again or continue to deform, a greater external force will be required or the environmental factors will be changed to make the dislocation overcome these obstacles. This is also the reason why the residual strain after the high stress plateau is smaller than before the plateau, which can be seen as equivalent to a slight increase in the strength of the rock salt after the high plateau and thus an increased ability to resist deformation.

Under a constant load, the creep-generated microhole defects gradually develop and continue to grow and then link with surrounding holes to form microcracks, which further link and converge to form the main crack through the crystal. In the later stage of the creep–fatigue test, when the creep-induced grain-piercing cracks meet the fatigue-induced cracks along the grain fractures or grain-piercing fracture cracks and converge, an interactive effect arises that accelerates the damage of the specimen. Therefore, the effect of creep–fatigue loading on the surrounding rock must be considered when evaluating the safety factor of CAES plants using salt caverns.

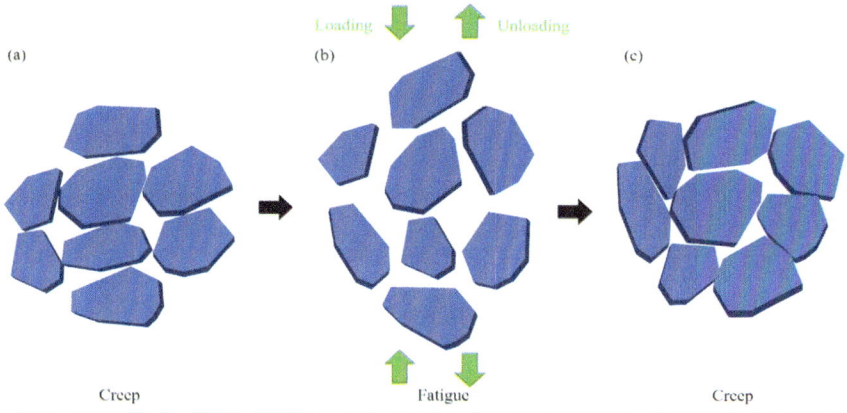

Fig. 5.17 Schematic diagram of the macroscopic level of the rock salt: **a** the last creep stage, **b** the cyclic loading stage after creep stage, and **c** the creep stage after cyclic loading

5.4 Conclusions

Due to the need for peak demand CAES power station in the change of gas pressure will exist in a certain time interval, that is, salt cavern reservoir surrounding rock is subjected to creep—fatigue alternating action. Therefore, when evaluating the stability of the salt cavern reservoir, the creep–fatigue mechanical properties of the rock salt need to be specifically studied. In this chapter, the creep–fatigue interaction of rock salt is investigated by setting different high stress plateaus durations (0, 5, 10, 15 and 20 min), and the reasons for the interaction are explained from a microscopic point of view. The main conclusions are as follows:

(1) The development law of creep–fatigue stress–strain curves is similar to that of conventional fatigue tests, and is also divided into three stages of combing-intensive-combining. The cyclic load has a promoting effect on creep deformation, but this promoting effect is gradually weakened with the increase of the high stress plateau time, and the creep deformation seems minimal when the high stress plateau is 15 min.
(2) The residual strain before the high stress plateau of the creep–fatigue test is greater than that after the high stress plateau, which is different from the results of the conventional rock salt fatigue test. With the increase of the time of the high stress plateau period, the difference of residual strain before and after the plateau increased, but this increase seems to saturate.
(3) The fatigue life of the salt-rock specimens decreases and the creep life increases with the increase of high-stress plateau duration. According to the creep–fatigue damage equation, it can be obtained that there is an interaction between creep and fatigue during the rock salt creep–fatigue test.
(4) During cyclic loading, the rearrangement of rock salt crystals accelerates the creep rate of rock salt during high stress plateaus. The cross-cutting and plugging of dislocations within the rock salt prevented the dislocations from continuing to slip, which in turn hardened the rock salt, resulting in a larger residual strain in the cycle before the high-stress plateau than in the cycle after the plateau.

References

1. Zongze Li, Fan Yang, Jinyang Fan, et al. Fatigue effects of discontinuous cyclic loading on the mechanical characteristics of sandstone[J]. Bulletin of Engineering Geology and the Environment, 2022, 81(8): 336.
2. JinYang Fan, Jie Chen, Deyi Jiang, et al. Fatigue properties of rock salt subjected to interval cyclic pressure[J]. International Journal of Fatigue, 2016, 90109–115.
3. Zengliang Gao, Yuxuan Song, Zhouxin Pan, et al. Nanoindentation investigation on the creep behavior of P92 steel weld joint after creep-fatigue loading[J]. International Journal of Fatigue, 2020, 134105506.
4. Stefan Holmström, Auerkari Pertti. A robust model for creep-fatigue life assessment[J]. Materials Science and Engineering: A, 2013, 559333–335.

References

5. I Salam, Tauqir A, Khan A-Q. Creep-fatigue failure of an aero engine turbine blades[J]. Engineering failure analysis, 2002, 9(3): 335–347.
6. Bara-Wasfi Al-Mistarehi, Khadaywi Taisir-S, Hussein Ahlam-Khaled. Investigating the effects on creep and fatigue behavior of asphalt mixtures with recycled materials as fillers[J]. Journal of King Saud University-Engineering Sciences, 2021, 33(5): 355–363.
7. Hongfu Liu, Xinyu Yang, Lijun Jiang, et al. Fatigue-creep damage interaction model of asphalt mixture under the semi-sine cycle loading[J]. Construction and Building Materials, 2020, 251119070.
8. RJH Thompson, Bonfield P-W, Dinwoodie J-M, et al. Fatigue and creep in chipboard: Part 3. The effect of frequency[J]. Wood science and technology, 1996, 30(5): 293–305.
9. Jie Zhang, Qiuhua Rao, Wei Yi. A New Creep–Fatigue Interaction Model for Predicting Deformation of Coarse-Grained Soil[J]. Materials, 2022, 15(11): 3904.
10. JunbaoWang, Qiang Zhang, Zhanping Song, et al. Microstructural variations and damage evolvement of salt rock under cyclic loading[J]. International Journal of Rock Mechanics and Mining Sciences, 2022, 152105078.
11. Zong-ze Li, Deyi Jiang, JinYang Fan, et al. Experimental study of triaxial interval fatigue of salt rock[J]. Rock and soil mechanics, 2020, 41(4): 1305–1312, 1322.
12. Anne-ES Forbes, Blake Steven, Tuffen Hugh. Entablature: fracture types and mechanisms[J]. Bulletin of Volcanology, 2014, 761-13.
13. Xiaohui Ni, Zhende Zhu, Xiaojuan Li, et al. Quantitative test study of meso-damage of rock under cyclic load[J]. Rock and soil mechanics, 2011, 32(7): 1991-1995.
14. Shangzhi Zhou, Jun Li, Ying Liu. Rock fatigue crack propagation mechanism based on far field cyclic compression[J]. Journal of Changsha University of Technology (Natural Science Edition), 2009, 6(1): 19-23.
15. Erik Rybacki, Niu Lu, Evans B. Semi-brittle deformation of Carrara marble: Hardening and twinning induced plasticity[J]. Journal of Geophysical Research: Solid Earth, 2021, 126(12): e2021JB022573.
16. Rashid-Geranmayeh Vaneghi, Thoeni Klaus, Dyskin Arcady-V, et al. Fatigue damage response of typical crystalline and granular rocks to uniaxial cyclic compression[J]. International Journal of Fatigue, 2020, 138105667.
17. J-Y Huang, JC E, Huang J-W, et al. Dynamic deformation and fracture of single crystal silicon: fracture modes, damage laws, and anisotropy[J]. Acta Materialia, 2016, 114136–145.
18. Taoying Liu, Cui Mengyuan, Li Qing, et al. Fracture and Damage Evolution of Multiple-Fractured Rock-like Material Subjected to Compression[J]. Materials, 2022, 15(12): 4326
19. Xiaoping Zhou, Xiaokang Pan, Berto Filippo. A state-of-the-art review on creep damage mechanics of rocks[J]. Fatigue & Fracture of Engineering Materials & Structures, 2022, 45(3): 627-652.
20. Dongxu Chen, LaiguiWang, Chuang Sun, et al. Particle flow study on the microscale effects and damage evolution of sandstone creep[J]. Computers and Geotechnics, 2023, 161105606.
21. BinxuWang, Tingchun Li, Qingwen Zhu, et al. Study on the creep properties and crack propagation behavior of single-fissure sandstone based on the damage bond model[J]. Theoretical and Applied Fracture Mechanics, 2023, 124103805.
22. Junbao Wang, Xinrong Liu, Zhanping Song, et al. A whole process creeping model of salt rock under uniaxial compression based on inverse S function[J]. Chinese Journal of Rock Mechanics and Engineering, 2018, 37(11): 2446-2459.
23. Shuang-Shuang Yuan, Qi-Zhi Zhu, Lun-Yang Zhao, et al. Micromechanical modelling of short-and long-term behavior of saturated quasi-brittle rocks[J]. Mechanics of Materials, 2020, 142103298.
24. Hongwen Jing, Qian Yin, Shengqi Yang, et al. Micro-mesoscopic creep damage evolution and failure mechanism of sandy mudstone[J]. International Journal of Geomechanics, 2021, 21(3): 4021010.
25. R-D Campbell. Creep/fatigue interaction correlation for 304 stainless steel subjected to strain-controlled cycling with hold times at peak strain[J]. ASME J. Eng. Ind, 1971, 93887–892.

26. K Gurumurthy, Srinivasan Balaji, Krishna Penchala-Sai, et al. Creep-fatigue design studies for process reactor components subjected to elevated temperature service as per ASME-NH[J]. Procedia Engineering, 2014, 86327–334.
27. Yukio Takahashi, Dogan Bilal, Gandy David. Systematic evaluation of creep-fatigue life prediction methods for various alloys[J]. Journal of Pressure Vessel Technology, 2013, 135(6): 61204.
28. L-K Severud, Winkel B-V. Elastic creep-fatigue evaluation for ASME code[A]//1987: 123–131.
29. Joseph Oldham, Abou-Hanna Jeries. A numerical investigation of creep-fatigue life prediction utilizing hysteresis energy as a damage parameter[J]. International journal of pressure vessels and piping, 2011, 88(4): 149-157.
30. Jill-K Wright, Carroll Laura-J, Sham T-L, et al. Determination of the creep-fatigue interaction diagram for Alloy 617[A]//American Society of Mechanical Engineers, 2016: V005T12A004.
31. R Lagneborg, Attermo RJMT. The effect of combined low-cycle fatigue and creep on the life of austenitic stainless steels[J]. Metallurgical Transactions, 1971, 21821–1827.
32. Junbao Wang, Qiang Zhang, Zhanping Song, et al. Experimental study on creep properties of salt rock under long-period cyclic loading[J]. International Journal of Fatigue, 2021, 143106009.
33. Deyi Jiang, Yao Cui, JinYang Fan, et al. Experimental study of mechanical characteristics of salt rock under discontinuous cyclic loading[J]. Rock and soil mechanics, 2017, 38(5): 1327–1334.
34. J-L Urai, Spiers C-J. The effect of grain boundary water on deformation mechanisms and rheology of rocksalt during long-term deformation. CRC Press, 2017: 149–158.
35. Guillaume Desbois, Urai Janos-L, de Bresser Johannes-HP. Fluid distribution in grain boundaries of natural fine-grained rock salt deformed at low differential stress (Qom Kuh salt fountain, central Iran): Implications for rheology and transport properties[J]. Journal of structural geology, 2012, 43128–143.
36. Tingting Xu, Arson Chloé. Self-consistent approach for modeling coupled elastic and viscoplastic processes induced by dislocation and pressure solution[J]. International Journal of Solids and Structures, 2022, 238111376.
37. C-J Spiers, Schutjens PMTM. Intergranular pressure solution in NaCl: Grain-to-grain contact experiments under the optical microscope[J]. Oil & Gas Science and Technology, 1999, 54(6): 729–750.
38. J-H Ter Heege, De Bresser JHP, Spiers C-J. Rheological behaviour of synthetic rocksalt: the interplay between water, dynamic recrystallization and deformation mechanisms[J]. Journal of Structural Geology, 2005, 27(6): 948–963.
39. JH-D Ter Heege, De Bresser JHP, Spiers C-J. Dynamic recrystallization of wet synthetic polycrystalline halite: dependence of grain size distribution on flow stress, temperature and strain[J]. Tectonophysics, 2005, 396(1–2): 35–57.
40. Hugh-C Heard. Steady-state flow in polycrystalline halite at pressure of 2 kilobars[J]. Geophysical monograph series, 1972, 16191–209.
41. G-M Pennock, Drury M-R. Low-angle subgrain misorientations in deformed NaCl[J]. Journal of microscopy, 2005, 217(2): 130–137.
42. G-M Pennock, Drury M-R, Peach C-J, et al. The influence of water on deformation microstructures and textures in synthetic NaCl measured using EBSD[J]. Journal of structural geology, 2006, 28(4): 588–601.
43. Jaroslav Pokluda, Černý Miroslav, Šandera Pavel, et al. Calculations of theoretical strength: State of the art and history[J]. Journal of computer-aided materials design, 2004, 111–28.
44. Stephen Horseman, Passaris Evan. Creep tests for storage cavity closure prediction[A]//1981: 119–157.
45. Xinrong Liu, Xin Yang, Junbao Wang. A nonlinear creep model of rock salt and its numerical implement in FLAC 3D[J]. Advances in Materials Science and Engineering, 2015, 2015.
46. Yujie Wei, Bower Allan-F, Gao Huajian. Recoverable creep deformation and transient local stress concentration due to heterogeneous grain-boundary diffusion and sliding in polycrystalline solids[J]. Journal of the Mechanics and Physics of Solids, 2008, 56(4): 1460-1483.

References

47. Michel Aubertin, Julien Michel-R, Servant Stéphane, et al. A rate-dependent model for the ductile behavior of salt rocks[J]. Canadian Geotechnical Journal, 1999, 36(4): 660-674.
48. P-D Portella, Rie K-T. Low cycle fatigue and elasto-plastic behaviour of materials[M]. Elsevier, 1998.
49. Yanfei Kang, JinYang Fan, Deyi Jiang, et al. Influence of geological and environmental factors on the reconsolidation behavior of fine granular salt[J]. Natural Resources Research, 2021, 30805–826.
50. D Grgic, Al Sahyouni F, Golfier F, et al. Evolution of gas permeability of rock salt under different loading conditions and implications on the underground hydrogen storage in salt caverns[J]. Rock Mechanics and Rock Engineering, 2022, 1–24.

Open Access This chapter is licensed under the terms of the Creative Commons Attribution-NonCommercial-NoDerivatives 4.0 International License (http://creativecommons.org/licenses/by-nc-nd/4.0/), which permits any noncommercial use, sharing, distribution and reproduction in any medium or format, as long as you give appropriate credit to the original author(s) and the source, provide a link to the Creative Commons license and indicate if you modified the licensed material. You do not have permission under this license to share adapted material derived from this chapter or parts of it.

The images or other third party material in this chapter are included in the chapter's Creative Commons license, unless indicated otherwise in a credit line to the material. If material is not included in the chapter's Creative Commons license and your intended use is not permitted by statutory regulation or exceeds the permitted use, you will need to obtain permission directly from the copyright holder.

Chapter 6
Creep–Fatigue Mechanical Characterization of Rock Salt Under Triaxial Stresses

Salt cavern reservoirs used in CAES power plants are often located deep underground, and the surrounding rock of the reservoir is in a triaxial stress state [1]. When the cavity is formed by water-soluble mining, the original stress balance of the reservoir surrounding rock is changed, and the differential stress in the stress field will cause continuous deformation of the surrounding rock. For all rock materials, the existence of confining pressure on the one hand increases the ultimate strength of the rock and improves the bearing capacity of the rock; on the other hand, it increases the toughness of the rock, so that some of the rocks in the shallow part of the performance of ordinary hard rocks, in the deep part of the performance of the large deformation of the soft rock characteristics. With the increase of the confining pressure, the rock will be transformed from brittle response to full ductile response, which is especially important for rock salts, and the temperature and confining pressure ranges in which the transformation occurs are much smaller than those of other types of rock materials [2]. The effect of the confining pressure on the ductility of rock salt is quite significant. Therefore, in order to be closer to the engineering reality and to reveal the creep–fatigue mechanical properties and damage evolution law of underground salt cavern surrounding rock, it is necessary to carry out triaxial rock salt creep–fatigue tests under different confining pressures on the same basis as creep–fatigue tests reported in Chap. 5.

6.1 Experimental Methods

6.1.1 Triaxial Creep–Fatigue Procedure

Geological conditions such as the general burial depth of the salt cavern storage and testing conditions such as the maximum capacity of the testing machine, were both taken into consideration. Four confining pressure (σ_c) values were considered: 3, 6, 9 and 12 MPa. Firstly, 12 rock salt samples were tested under monotonic compressive loading until failure, at different σ_c. Each σ_c level was repeated three times, and the average triaxial compressive strength (TCS) are given in Table 6.1.

According to the average TCS for different σ_c, the stress level for triaxial creep–fatigue tests were designed. For each confining pressure, the upper stress $\sigma_{a,\,max}$ is taken equal to 85% of the peak differential stress, and the lower stress $\sigma_{a,\,min}$ is set to 3% of the peak differential stress, as given by the following equation:

$$\begin{cases} \sigma_{a,max}=0.85(\sigma_{TCS}-\sigma_c)+\sigma_c \\ \sigma_{a,min}=0.03(\sigma_{TCS}-\sigma_c)+\sigma_c \end{cases} \quad (6.1)$$

The values for upper and lower stresses are presented in Table 6.2. Note that the plateau duration is fixed to 5 min in this test series.

The tests are defined as triaxial creep–fatigue (TCF) tests. For the TCF tests, the ratio of the upper to lower stress bounds is the same, despite different σ_c. The specific loading and unloading paths are as follows: Initially, using a computer-controlled program, the σ_c is increased to a predetermined level. Subsequently, the axial stress is increased at the loading rate of 2 kN/s up to $\sigma_{a,max}$, where it is also held T-stage. This step is carried out to simulate the rheological condition of the surrounding rock of the salt cavern after the initial injection of air owing to the sustained high-pressure environment, consistent with the purpose of the uniaxial creep fatigue test. This T-stage is then followed by a decrease in stress to $\sigma_{a,min}$ at the same 2 kN/s rate. After, linear loading and unloading are applied between $\sigma_{a,min}$ and $\sigma_{a,max}$ still at a rate of 2 kN/s, and a high stress plateau of duration $\Delta t = 5$ min is applied every two cycles, as shown in Fig. 6.1. The testing utilized a higher maximum stress level, $\sigma_{a,max}$, and

Table 6.1 Compressive strengths of the rock salt samples under various confining pressure

Confining pressure/MPa	Test No. 1	Test No.2	Test No.3	Average TCS/MPa
0	29.5	30.2	30.6	30.1
3	54.3	57.5	61.6	57.8
6	71.1	78.2	73.3	74.2
9	86.8	84.7	90.4	87.3
12	102.5	108.6	100.3	103.8

Table 6.2 Parameters adopted in the TCFT

Confining pressure/MPa	Upper stress bound/MPa	Lower stress bound/MPa	Plateau time/min
3	49.2	4.56	5
6	64.1	8.2	5
9	76.2	11.6	5
12	90.5	14.8	5

6.1 Experimental Methods

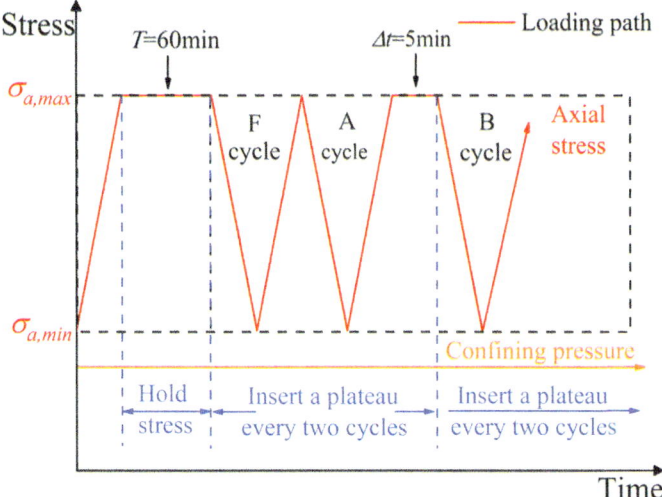

Fig. 6.1 Loading paths during the triaxial creep–fatigue (TCF) tests

a lower minimum stress level, $\sigma_{a,min}$, than actual operating pressure of the CAES salt cavern (Minimum air pressure 10 MPa, Maximum air pressure 25 MPa). This was primarily done to minimize the duration of the testing, enabling a full test to be conducted within a single workday.

The term "A cycle" refers to the cycle before the high-stress plateaus and "B cycle" to the cycle after the high-stress plateaus. The first cycle (F) belongs to is neither a pre-plateau nor a post-plateau and is called the first cycle. As for uniaxial creep–fatigue tests presented in Chap. 2, the purpose of this setup is to compare and analyze the impacts of two different stress cyclic loadings on the mechanical behaviors of rock salts in the same samples and to avoid specimen variability.

6.1.2 Test Procedure

(1) Installation of test pieces, swabbing the end face of the upper and lower pressure head, the test piece is placed between the upper and lower pressure head, so that the center of the three into a straight line; and then the test piece and the sting block set on the heat shrinkable oil-proof sleeve and blowing tightly, with adjustable stainless steel hoop to close the test piece. Lower the triaxial chamber, screw the triaxial chamber fixing screws and top closing screws.

(2) Start the MTS815 rock mechanics test system and apply 0.1 kN–0.5 kN pressure to fix the specimen. Then open the exhaust valve of the pressure chamber, start the peripheral pressure oil pump, inject oil into the pressure chamber, exhaust

the air while injecting oil, and tighten the exhaust valve when the air in the pressure chamber is exhausted.

(3) loading: 0.05 MPa/s loading speed to apply pressure to a predetermined pressure value, in the test process so that this side of the pressure value remains constant, the range of change should not exceed the selected value of ± 2%. Then the computer-controlled program is used to apply a predetermined load until the specimen is destroyed. Subsequently, the pressure is unloaded and the specimen is removed and the damage condition is recorded.

6.2 Test Results and Analysis

6.2.1 Stress–Strain Curve in the Triaxial Creep–Fatigue Tests

The TCF tests results under various confining pressure σ_c are shown in Fig. 6.2. Taking the result under $\sigma_c = 3$ MPa as an example, it is observed that, after initial loading causing a deformation, the high-stress plateau lasting one hour (T-stage) causes significant deformation as well. During the subsequent alternating creep–fatigue stage, the experimental curve exhibits, again, three distinct deformation stages of sparse-dense-sparse, namely the decelerating deformation stage (I), the stable deformation stage (II), and the accelerating deformation stage (III), illustrated in Fig. 6.2a. This indicates that the TCF test follows the characteristics of a conventional fatigue test [3]. This TCFT is also very similar to the pure fatigue test (no upper stress plateau) under triaxial loading. However, the significant difference is that the proportion of II stage for the stress–strain curve are more obvious of a pure fatigue test under triaxial loading [4]. Stage I of the stress–strain curve is significantly condensed in the TCF test, as it is in the creep–fatigue test with uniaxial loading. Stages II and III are more prominent. These results exhibit the characteristic of three-stage in the TCF test with σ_c changes. Firstly, the deformation of stage I shortens, because the rock salt specimens are compacted and densified during the one-hour high-stress plateau. Secondly, as the σ_c increases, the densification effect of the σ_c leads to the fusion of stage I and stage II of the experimental curve, and the proportion of deformation in Stage II to the total deformation increases. Thirdly, the presence of stage III rather than stage I is more evident during the test with σ_c are 3, 6, 9 MPa, reflecting the change in the high-stress plateau from a compaction/density effect to the force driving the specimen destruction during the final damage phase of the creep–fatigue test. Finally, When the σ_c rises to 12 MPa, the transition from II to the III stage also becomes unclear, the whole stress–strain curve shows the characteristics of only one stage.

To reflect the impact of the σ_c on the creep–fatigue deformation of rock salts, the total strain of the test specimen is defined as ε_t and the strain of the first loading of the specimen to the upper stress is defined as ε_l (see Fig. 6.2b). The strain after the first loading, $\Delta\varepsilon$, can be calculated as:

6.2 Test Results and Analysis

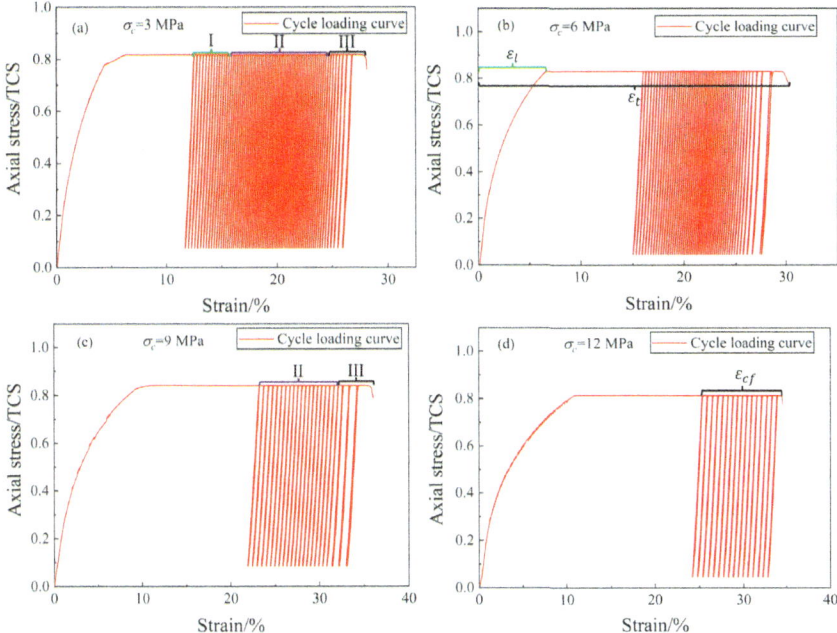

Fig. 6.2 Stress–strain curves for TCF tests with confining pressure values of **a** 3 MPa, **b** 6 MPa, **c** 9 MPa and **d** 12 MPa

$$\Delta \varepsilon = \varepsilon_t - \varepsilon_l \tag{6.2}$$

According to Fig. 6.3, with increasing σ_c, ε_t shows an increasing trend. This is because as σ_c increases, the loose rock salt particles originally damaged by the stress are compacted tightly, the internal fissures are closed more completely, the fissure surfaces that can produce slip are locked, the self-healing effect is enhanced and the structure of the rock salt is more solid [5]. A similar trend is shown for $\Delta \varepsilon$. This reflects that the rock salt behavior gradually changes from brittle–ductile to ductile with the σ_c increase. This increasing plastic deformation capacity, reported extensively for monotonic loading, is also observed here under cyclic loading.

The fatigue and creep lifes of rock salt specimens under different σ_c, shown in Fig. 6.4, are found to decrease with confining pressure increasing. This is not exactly consistent with the triaxial pure fatigue test [6]. Our results can be explained by the influence of two types: Firstly, the addition of creep results in a greater deformation during the creep period at higher upper stress limits, causing a reduction in the percentage of plastic deformation generated by fatigue and a decrease in fatigue life. Secondly, the impact of the one-hour high stress plateaus following the initial loading on the creep deformation of rock salt is non-linear. This leads to a decrease in the proportion of alternating creep–fatigue deformation stages to the total strain

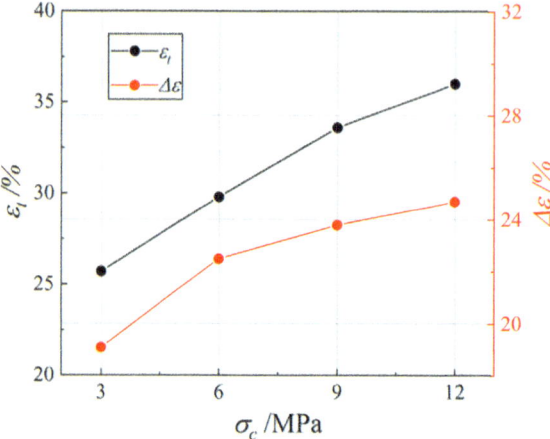

Fig. 6.3 The total strain before failure (ε_t) and strain after first loading ($\Delta\varepsilon$) for TCF tests under different confining pressure values

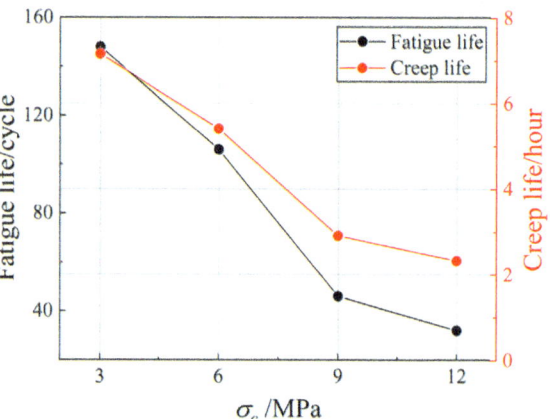

Fig. 6.4 Fatigue life and creep life for TCF tests under different confining pressure values

in high-confining pressure test samples. The implications of this outcome for the construction of salt cavern reservoirs will be discussed in Sect. 6.4.

6.2.2 Impact of Confining Pressure on Creep Deformation in Rock Salt in the Triaxial Creep–Fatigue Tests

For triaxial creep–fatigue tests, the creep rate and deformation are more important because the brittle-ductile response due to confining pressure is more pronounced for soft rocks. The same Eq. (5.2) as in Sect. 5.2.2 is used for the calculation of creep rate v_c.

Taking the TCF test under 3 MPa for analysis, the test results are shown in Fig. 6.5. First, it can be found that the curve is very similar to the curve obtained for uniaxial

6.2 Test Results and Analysis

creep–fatigue test. During the initial T stage, the v_c of the rock salts show a nearly linear decreasing tendency, but the rate of decrease slows. This indicates that at the higher stress level, the interior structure of the rock salt undergoes rapid adjustment, a large number of fractures are filled and closed, a hardening phenomenon occurs, and the resistance to deformation is enhanced. While entering the creep–fatigue stage, after two cycles of loading, v_c shows a jump, followed by a slow decline into the stable deformation stage. During this stable stage, v_c remains basically unchanged.

Figure 6.6 reflects the variation in the creep rate in the initial one-hour high-stress phase of the rock salts under different σ_c. The creep rate development patterns of the four tests are similar, but the creep rate at larger confining pressure values is much larger than the creep rate at lower confining pressure values. In addition, the distribution of the creep rate also shows obvious zoning characteristics; that is, the creep rates of the samples under σ_c of 12 and 9 MPa (within the red dashed circles) are significantly larger than those of 6 and 3 MPa (within the green circles). The reasons for the above phenomenon are as follow: One reason is that the increase in the upper stress plateau (with the confining pressure) causes the creep rate to increase. Another reason concerns the brittle–ductile transformation for the rock salts [7]. It has been pointed out that the σ_c has a significant impact in the microfracture and shear expansion of rock salt [8]. The plastic deformation at a low σ_c is dominated by crack extension, while the plastic deformation at a high σ_c is dominated by the dislocation mechanism. With the σ_c increases, the rock salt shows super mobility and significant ductile characteristics [9]. The deformation characteristics of the rock salt are no longer obvious with the change in the σ_c. In this work, we speculated that

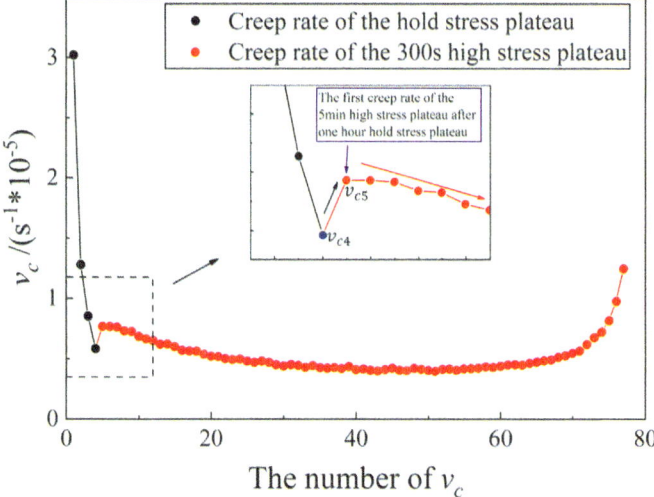

Fig. 6.5 Creep rate (v_c) for TCF test at 3 MPa confining pressure. The black dot is the v_c at the one-hour high-stress plateau, and the red dot is the v_c at the high-stress plateau during the fatigue creep stage

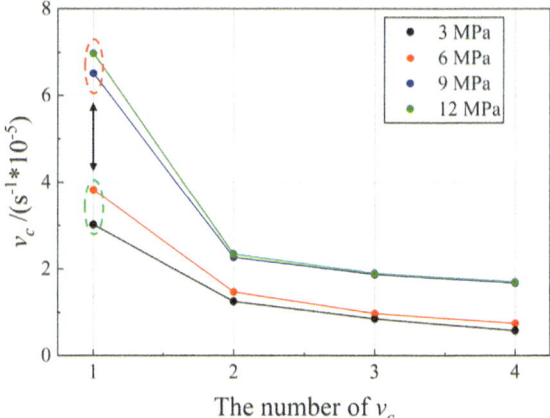

Fig. 6.6 The trends of the first four v_c values during the T stage under different confining pressure

6–9 MPa is the critical σ_c for the brittle–ductile transformation of rock salts, which should be verified by more tests in the future.

The incorporation of fatigue was able to enhance the v_c of the rock salts during the high-stress plateaus stage, which was demonstrated in all four sets of tests. However, the effect of different σ_c on the increase in v_c differs. For this purpose, we calculated the ratio of the last v_c datapoint of the first T stage to the first v_c datapoint in the creep–fatigue stage for the rock salt separately and defined it as the growth rate β defined as:

$$\beta = \frac{v_{c5}}{v_{c4}} \qquad (6.3)$$

where v_{c5} and v_{c4} indicate the 5th and 4th creep rates in the different TCF tests, respectively, as shown in the detail of Fig. 6.5. The statistical result with different σ_c is shown in Fig. 6.7. The growth rate tends to decrease linearly with increasing σ_c. The reason why cyclic loading elevates the creep rate is that the unloading and loading of pressure opens the otherwise closed fracture surfaces and continues to produce slip at the next high-stress plateau. The presence of the σ_c inhibits this behavior, and this constraint becomes more pronounced as the σ_c increases. Similar to the reason for the decrease in residual strain difference, this effect will not change when the confining pressure reaches a threshold. Note that the value of this growth rate is influenced by the choice to split the first T stage into 4 segments. However, it was verified that the growth rate decreases linearly as a function of confining pressure, whatever the chosen number of segments.

To illustrate the impact of different σ_c on creep deformation at a given time, we calculated the average creep strain $\bar{\varepsilon}_c$ of the rock salt during second stage as:

$$\bar{\varepsilon}_c = \frac{\sum_j^k \varepsilon_{ci}}{k - j} \qquad (6.4)$$

6.2 Test Results and Analysis

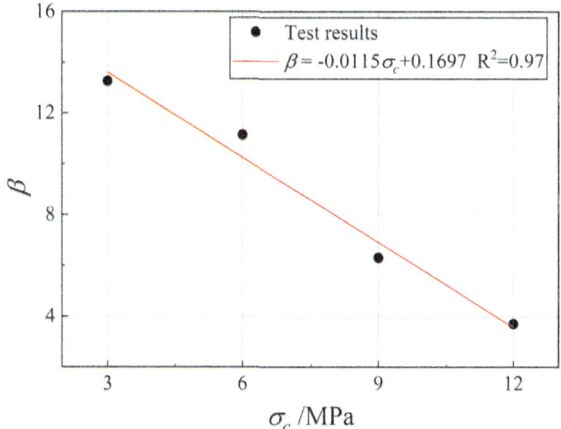

Fig. 6.7 Ratio beta of the last v_c of the first T stage and the first v_c of the creep–fatigue stage in four TCF tests

where j and k indicate the beginning and end loading cycle numbers of the second stage in every TCFT. The results are plotted in Fig. 6.8.

As the σ_c rises, the $\bar{\varepsilon}_c$ of the rock salts increases significantly in the high-stress plateaus stage. The $\bar{\varepsilon}_c$ of the 12 MPa test is nearly five times higher than that of the 3 MPa test. This is consistent with the evolutionary model on the stress–strain curve in a TCF test.

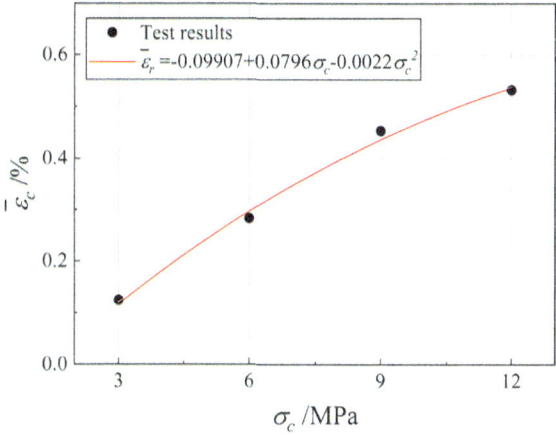

Fig. 6.8 Average creep strain during the high-stress plateau in the stable deformation stage under different confining pressures

6.2.3 Impact of Confining Pressure on Residual Strain in Rock Salts During the Triaxial Creep–Fatigue Tests

In order to reveal the effect of confining pressure on the triaxial creep–fatigue test of rock salt, the residual strains of rock salt specimens under four confining pressure values, calculated as in Sect. 5.2.3, are shown in Fig. 6.9.

The residual strain ε_r of the TCF tests on rock salt under 3 and 9 MPa σ_c were selected for analysis. We find that the ε_r distribution is not the same as for conventional fatigue test. The two confining pressure values present the same characteristics, i.e., larger values of ε_r in cycle A than in cycle B, which is consistent with the conclusion obtained from the uniaxial creep–fatigue tests presented in Chap. 5. The same phenomenon was observed for the confining pressure values of 6 and 12 MPa TCF tests. This indicates that the presence of the σ_c does not change this phenomenon. The reason for this phenomenon is the compaction of the rock salts crystals under the high-stress plateaus and the accumulation of material filling the internal cracks. It can be assumed that hardening occurs during the creep stage, causing the specimen to resist deformation more. Thus, the compressive strength of the rock salts is raised.

Comparing the ε_r of the tests at σ_c of 3 and 9 MPa, the apparent difference is that the ε_r of the 3 MPa test still exhibits a clear decelerating deformation stage. In the alternating creep–fatigue phase, stage I slowly decreases to the stable II stage. The ε_r in the 9 MPa test goes directly to stage II, followed by rapid destruction in the accelerating deformation stage. This is similar to the pattern exhibited by the creep–fatigue stress–strain curves of rock salts, as discussed on the Sect. 6.2.1.

In addition, the difference in ε_r between cycle A and cycle B varies significantly with σ_c. The difference in ε_r for 9 MPa is less important than for 3 MPa. To more accurately compare the effects of different σ_c on ε_r before and after the high-stress plateau, the average residual strain difference between cycle A and cycle B, $\Delta\bar{\varepsilon}_r$, during the stable deformation stage (stage II was computed by Eqs. (4.1, 4.2, 5.5) defined in Sects. 4.2.2 and 5.2.3. Figure 6.10 shows the calculated results.

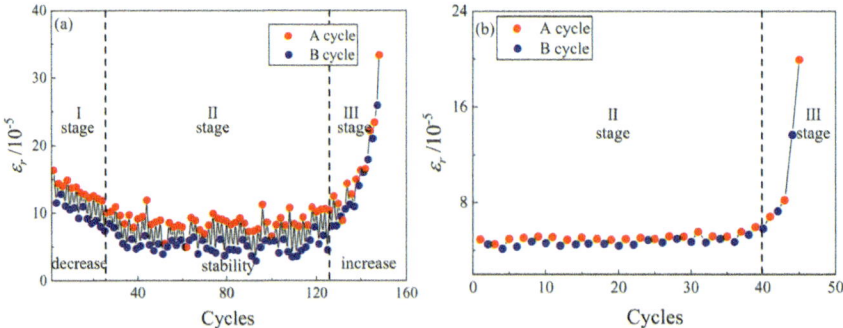

Fig. 6.9 Axial residual strain (ε_r) at each cycle for TCF tests at confining pressure values of **a** 3 MPa and **b** 9 MPa

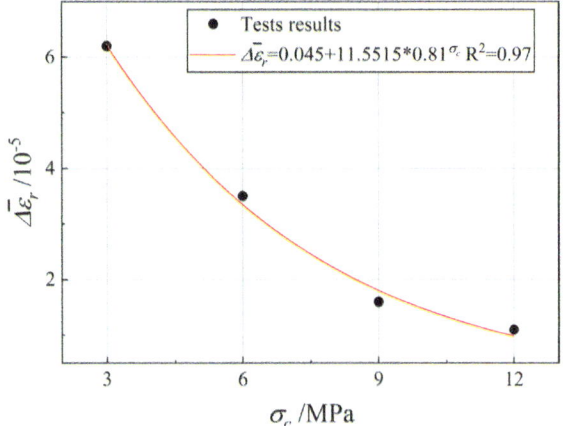

Fig. 6.10 Average residual strain difference between cycles A and B ($\Delta\bar{\varepsilon}_r$) for TCF tests under different confining pressure values

$\Delta\bar{\varepsilon}_r$ exhibits a declining exponential trend with rising σ_c and the rate of decrease slows. Because as the σ_c rises, the upper stress bound increases with the same stress ratio. The rock salt is compacted more tightly during the creep stage. As the test proceeds, the fractures inside the rock salt will completely close during the high-stress plateau. The initial defects and cracks have difficulty further developing. The hardening effect in the creep stage is no longer obvious, and $\Delta\bar{\varepsilon}_r$ decreases. The decreasing rate indicates that the development and expansion of microcracks are suppressed after the increase in the σ_c reach a certain level. The high σ_c prevents relative slip between the grains of the rock salt. As the increase in the σ_c, it is difficult for damage to evolve, even if the resulting plastic flow produces large deformation [10]. The impact of the σ_c in the creep–fatigue behavior of the rock salts reaches a threshold.

6.3 Analysis of Confining Pressure Effects on Creep–Fatigue Properties of Rock Salts

6.3.1 Influence of the Increase of the Confining Pressure on the Transformation of Brittle Ductility in Rock Salts

Considering the three-dimensional stress states of underground geotechnical engineering, the research of the mechanical properties of rocks under triaxial stress can reflect the actual loading of underground works more realistically. Researchers have also conducted many triaxial mechanical tests on rocks, and this approach is especially important for rock salt. In contrast to the pure fatigue test on rock salt under triaxial stress, the introduction of a high stress plateau causes hardening of the rock salts during the creep phase, resulting in a lower ε_r for the B cycle following the high-stress plateaus than the A cycle before it. At the same time, the introduction of cyclic loading changes the inherent arrangement of rock salt crystals on the high-stress plateaus during unloading and reloading of stress, leading to crystal disorder. As the differences in crystal orientation between crystals increase, the v_c also increases. This is consistent with the reasons for the above phenomena observed in uniaxial creep–fatigue tests. The uniqueness of triaxial testing lies in the fact that with increasing σ_c, both the ε_r difference between the fatigue cycles before and after the high-stress plateaus and the enhancement impact of fatigue on the v_c are reduced. This is because the σ_c inhibits the expansion of fatigue cracks and the slip of rock salt crystals, and this effect becomes more significant with increasing σ_c.

All the rock behavior transforms from a brittle response to a full ductile response with the σ_c increases. However, the transformation pressure of rock salts (6–9 MPa) is lower than that of other rock materials [2, 11]. Under uniaxial compression, rock salts exhibit certain brittle characteristics, and damage occurs within a small strain range. At this time, the cracks of rock salts are related to mainly tensile damage, and the cracks will develop along the lower-strength grain boundaries [12]. With increasing axial stress, cracks along the macroscopic cleavage along the direction of the maximum principal stress are formed. At low σ_c (3 MPa for Fig. 6.11a and 6 MPa for Fig. 6.11b), the plastic zone of the rock salt is extended, and there is a plastic flow phenomenon. Finally, the specimen is sheared along a single oblique plane of damage. However, at σ_c of 9 MPa (Fig. 6.11c), the rock salt exhibited stronger flow. On the surface of that damaged specimen, X-shaped conjugate shear zones were observed, but none of them were significantly expanded yet. This indicates that as the increase in the σ_c, the initiation of damage to the specimen is gradually delayed, the accumulated crack volume and the accumulated damage caused by shear cracks increase, and the development of cracks becomes increasingly difficult [13]. We speculate that 6–9 MPa can be regarded as the critical σ_c for the brittle-ductile transformation of rock salts. When the σ_c continues to rise to 12 MPa (Fig. 6.11d), the final damage morphology of the rock salt specimen shows a short, thick cylindrical shape, similar to a drum, with only a few cracks hardly visible.

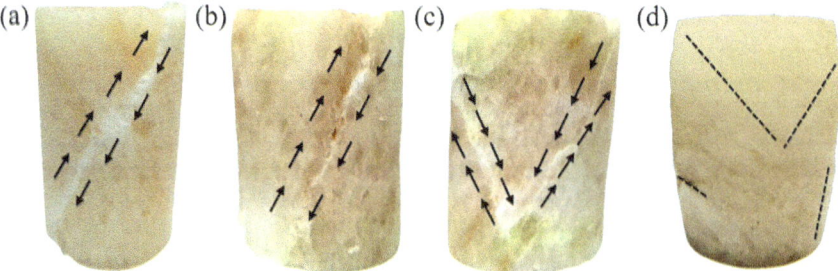

Fig. 6.11 Photos of rock salt specimens after testing under different confining pressure values **a** 3 MPa, **b** 6 MPa, **c** 9 MPa, **d** 12 MPa

6.3 Analysis of Confining Pressure Effects on Creep–Fatigue Properties ...

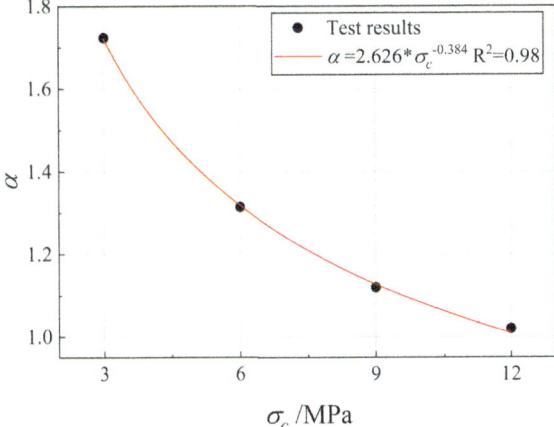

Fig. 6.12 Pressure density factor alpha under different confining pressure values

However, there is a lack of quantitative indicators for evaluating and judging the degree and state of the ductile deformation of rock salts. We surmise that the ratio of creep strain for the first high-stress plateau after two cycles of loading and mean creep strain of the middle 5% of the cycles during the creep–fatigue stage, i.e., the compression density coefficient (α), can be used to illustrate the degree of ductile deformation of the rock salts in this test. The compression density coefficient results are shown in Fig. 6.12; here α shows a decreasing trend with increasing σ_c and is reduced to approximately 1.07 under 12 MPa of σ_c. We believe that after reaching a certain σ_c, the internal fractures in the rock salt are likely to close completely; thus, the initial defects have difficulty developing. It is difficult for the fatigue cycle load to cause the rock salt fracture surfaces to slip again, and the rock salt specimen truly becomes a uniformly dense rock sample at the macroscopic scale.

Another point of interest is that the total strain ε_t and strain after first loading $\Delta \varepsilon$ of the rock salt indicates an increasing trend as σ_c rises. Nevertheless, the percentage of strain in the creep–fatigue alternating phase (ε_{cf}) to total strain (ε_t), shown in Fig. 3.2d, defined as γ indicates a decreasing trend. As shown in Fig. 6.13, γ under a low σ_c is still above 50%, but γ under 12 MPa of σ_c is only 32.1%. This indicates that the high stress has a greater impact in the deformation for the rock salts at the T stage under equal stress ratio conditions.

The effect of increasing the confining pressure on the one hand increases the ultimate strength of the rock and improves the bearing capacity of the rock; on the other hand, it increases the toughness of the rock, so that some of the rocks, which behave as ordinary hard rocks in the shallow part of the rock, show the characteristics of large deformation of the soft rock in the deep part of the rock. From the results of the compression test, it can be seen that the Young's modulus and deformation capacity of the rock salt become larger with the increase of the confining pressure. Because the confining pressure can limit the generation and expansion of rock salt cracks and prevent the relative slip between rock salt grains, the strength and ductile deformation capacity of rock salt are enhanced. But with the increase of the confining pressure,

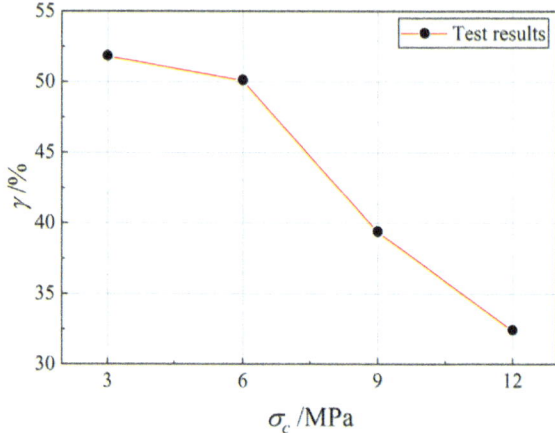

Fig. 6.13 Strain ratio gamma of alternating creep–fatigue stages to all the stress–strain curve stages for different confining pressure values

this effect of the confining pressure is getting smaller and smaller. Therefore, it is possible to reach a certain confining pressure, rock salt internal cracks completely closed, the original defects are difficult to develop, rock salt specimens really become macroscopically homogeneous and dense rock samples, and the same as found by other researchers, to get Young's modulus has nothing to do with the confining pressure of the conclusion. Because Young's modulus (elastic modulus) as an inherent property of rocks should not vary with the confining pressure. However, due to the presence of natural rock defects, the elastic modulus may show an increasing trend with increasing confining pressure. But when the confining pressure reaches a certain value, the internal primary cracks and defects in the rock will close. The rock then becomes an ideal state, appearing macroscopically as a uniformly dense rock, and at this point, Young's modulus becomes independent of the confining pressure [14].

6.3.2 Mechanical Interpretation of Rock Salts Affected by Confining Pressure and the Effect of Burial Depth on the Deformation of Surrounding Rock of the Salt Cavern

The brittle fracture evolution is very important for understanding the damage mechanism of rocks, and a large amount of literature has been published to study the fracture process of rocks. The results show that the confining pressure plays an important role in the damage and destruction of brittle rocks. The peak strength, damage evolution and damage pattern of specimens depend on the confining pressure. Moreover, the compressive strength of rocks increases in a nonlinear manner with increasing the confining pressure [15].

The reasons for the deformation affected by confining pressure are explained below in terms of theoretical mechanics. The stress tensor at a point can be

6.3 Analysis of Confining Pressure Effects on Creep–Fatigue Properties ...

decomposed into two parts: the spherical tensor part and the bias tensor part:

$$\begin{bmatrix} \sigma_{11} & \sigma_{12} & \sigma_{13} \\ \sigma_{21} & \sigma_{22} & \sigma_{23} \\ \sigma_{31} & \sigma_{32} & \sigma_{33} \end{bmatrix} = \begin{bmatrix} \sigma_0 & 0 & 0 \\ 0 & \sigma_0 & 0 \\ 0 & 0 & \sigma_0 \end{bmatrix} + \begin{bmatrix} \sigma_{11}-\sigma_0 & \sigma_{12} & \sigma_{13} \\ \sigma_{21} & \sigma_{22}-\sigma_0 & \sigma_{23} \\ \sigma_{31} & \sigma_{32} & \sigma_{33}-\sigma_0 \end{bmatrix} \quad (6.5)$$

In the above equation

$$\sigma_0 = \frac{1}{3}(\sigma_{11} + \sigma_{22} + \sigma_{33}) = \frac{I_1}{3} \quad (6.6)$$

where I_1 is the first invariant of the stress tensor and represents the average stress. Similarly, the strain tensor can be decomposed into the sum of the spherical tensor and the bias tensor:

$$\begin{bmatrix} \varepsilon_{11} & \varepsilon_{12} & \varepsilon_{13} \\ \varepsilon_{21} & \varepsilon_{22} & \varepsilon_{23} \\ \varepsilon_{31} & \varepsilon_{32} & \varepsilon_{33} \end{bmatrix} = \begin{bmatrix} \varepsilon_0 & 0 & 0 \\ 0 & \varepsilon_0 & 0 \\ 0 & 0 & \varepsilon_0 \end{bmatrix} + \begin{bmatrix} \varepsilon_{11}-\varepsilon_0 & \varepsilon_{12} & \varepsilon_{13} \\ \varepsilon_{21} & \varepsilon_{22}-\varepsilon_0 & \varepsilon_{23} \\ \varepsilon_{31} & \varepsilon_{32} & \varepsilon_{33}-\varepsilon_0 \end{bmatrix} \quad (6.7)$$

In the equation,

$$\varepsilon_0 = \frac{1}{3}(\varepsilon_{11} + \varepsilon_{22} + \varepsilon_{33}) = \frac{\theta}{3} \quad (6.8)$$

where θ is the volume strain.

The decomposition of the stress or strain tensors into two parts, the spherical tensor and the bias tensor (or deviatoric tensor), is of profound physical content. The deformation of a point within a rock can be divided into two parts: volume strain and shape change, with the spherical tensor part of the strain tensor representing its volume change and the partial tensor part corresponding to its shape change.

In order to study the change in the volume of the rock, three sets of elastic stress–strain relations are first written out:

$$\begin{aligned} \sigma_{22} &= \lambda\theta + 2\mu\varepsilon_{22} \\ \sigma_{22} &= \lambda\theta + 2\mu\varepsilon_{22} \\ \sigma_{33} &= \lambda\theta + 2\mu\varepsilon_{33} \end{aligned} \quad (6.9)$$

Adding the three equations above and dividing by 3 yields:

$$\begin{bmatrix} \sigma_0 & 0 & 0 \\ 0 & \sigma_0 & 0 \\ 0 & 0 & \sigma_0 \end{bmatrix} = 3k \begin{bmatrix} \varepsilon_0 & 0 & 0 \\ 0 & \varepsilon_0 & 0 \\ 0 & 0 & \varepsilon_0 \end{bmatrix} = k \begin{bmatrix} \theta & 0 & 0 \\ 0 & \theta & 0 \\ 0 & 0 & \theta \end{bmatrix} \quad (6.10)$$

In the equation,

$$k = \frac{2\lambda + 2\mu}{3} \tag{6.11}$$

where k is the volumetric compression modulus of the rock, and from the above Eq. (6.10) it is clear that the volumetric strain theta of the rock is related to σ_0 and only to σ_0 (this is clearer from the fact that the sum of the main diagonal elements of the bias tensor is always equal to zero)

For intact rocks, fracture is the main form of damage that occurs. Rupture occurs as soon as the pressure σ_1 to which the rock is subjected exceeds its strength. The rupture criterion when the rock is in the stress state of $(\sigma_1, \sigma_2, \sigma_3)$:

$$\sigma_1 = f(\sigma_2, \sigma_3) \quad \sigma_1 \geq \sigma_2 \geq \sigma_3 \tag{6.12}$$

The above equation is the condition under which rupture occurs, at which point σ_1 is also called the strength under the given σ_2, σ_3 conditions. In fact, it is also the relationship between the strength σ_1 and σ_2, σ_3.

If only the simplest case is considered, we usually use the linear Coulomb rupture criterion. Coulomb's law expressed in terms of positive stresses σ and tangential stresses τ on the rupture surface is

$$|\tau| = S_0 + \mu\sigma \tag{6.13}$$

where S_0 and μ are material constants related to the rock type. S_0 is called the strength of aggregation, often referred to as cohesion in engineering; μ is called the coefficient of internal friction. Another expression for the Coulomb criterion is obtained when expressed in terms of principal stresses σ_1 and σ_3:

$$\sigma_1 = C_0 + \xi\sigma_3 \tag{6.14}$$

where C_0 is a constant, for the uniaxial compressive strength of the rock, ξ is the coefficient of the influence of the confining pressure on the strength of the rock, which indicates that when σ_3 increased by a factor of 1, so that the rock rupture the required value of σ_1 needs to be increased by $\xi\sigma_3$ times, the parameters of the S_0, μ and the relationship between C_0 and ξ follows

$$C_0 = 2S_0\left[\left(\mu^2 + 1\right)^{\frac{1}{2}} + \mu\right] \tag{6.15}$$

$$\xi = \left[\left(\mu^2 + 1\right)^{\frac{1}{2}} + \mu\right]^2 \tag{6.16}$$

For most rocks, $\mu \approx 0.6 - 1.0$, $S_0 = 150 - 360$ KPa. Thus, an increase in the confining pressure σ_3 greatly increases the rupture strength σ_1. Figure 6.14 shows the difference between the two expressions of strength for rocks, as shown in Eqs. (6.13) and (6.14).

Fig. 6.14 Two kinds of expression for rock rupture criterion

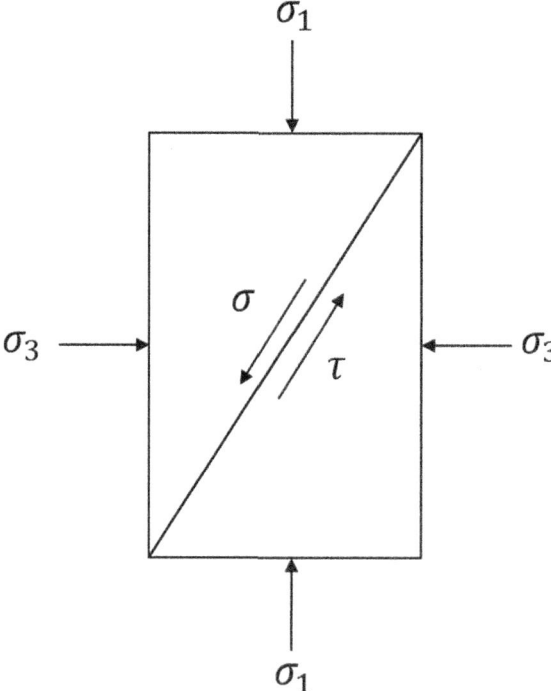

Before summarizing the implications of the test results on the locating, construction, and operation of a CAES salt caverns, it is necessary to explain the distribution pattern of ground stress. In 1978, E.T. Brown et al. statistically analyzed 120 representative crustal stress (σ_v) measurements collected worldwide and found that the measured values of ground stress are generally compressive stresses and that very few are tensile stresses, and they increase with increasing depth (H) [16]. The linear correlation between stress and depth was strong in the depth range of 2500 m. The vertical stress could be calculated as:

$$\sigma_v = 0.027H \tag{6.17}$$

In addition, the ratio of the average horizontal stress to the vertical stress ranges from 0.5 to 3.5. Considering that the vast majority of salt caverns are within 2000 m of the ground surface, Eq. (6.17) can be used to estimate the surrounding rock stresses in salt caverns. According to the above test result, the recommendations are as follow: For a horizontal dissolution cavity (Fig. 6.15a), the height (h_4) of the salt cavern in the vertical direction does not varies much. When a CAES plant is operating, the operation plan of the salt cavern can be adjusted according to the depth (h_3). For salt caverns with depths greater than 500 m, the creep–fatigue alternating load caused by the change in the frequency of peak regulation has less impact on the surrounding rock. The frequency of peaking can be increased appropriately to improve the efficiency of the CAES plant. At shallow depths, the impact of creep–fatigue interaction on the safety operation factor must be considered. Regarding the vertical dissolution cavity (Fig. 6.15b), a salt cavern has a large height (h_2) variation in the vertical direction. In addition to the general area of weakness in the surrounding rock–the roof of the salt cavern–we should also pay attention to the differences in the mechanical characteristics of the surrounding rocks due to the variation in depth (h_1). Take the example of the Hutchinson salt mine in Kansas, USA [17]. If h_1 of the salt cavern is 150 m and h_2 is 200 m, according to Eq. (6.17), the surrounding rocks above (e.g., yellow dashed box) and below (e.g., red dashed line) are under approximately 4 and 9 MPa of σ_c, respectively. The brittle and ductile transformation characteristics of the rock salt should be considered when designing the gas pressure. Otherwise, the salt cavern may not reach its service life and fail.

Moreover, the effect of the T stage and S_l on rock salt creep–fatigue behavior suggests that we should pay attention to the timing of dissolution cavity construction and the gas pressure and duration of the first gas injection when utilizing and modifying old cavities for the construction of CAES facilities, as these factors are equally important for the stability of the CAES salt cavern.

There are some deviations between experimental design and engineering practice due to the limited experimental conditions, such as load frequency, stress level and temperature condition. The crustal stress conditions in which the corresponding underground works are located are also extremely complex. Despite these limitations, this work can still help engineers evaluate the stability of CAES salt caverns.

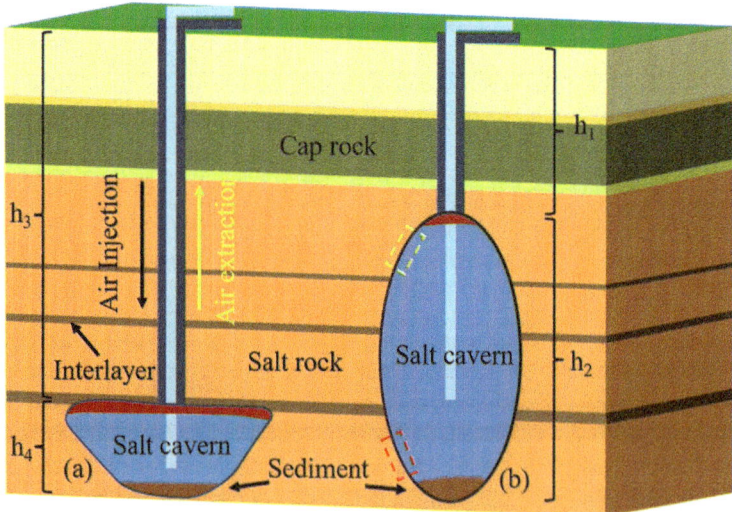

Fig. 6.15 Schematic diagram of two salt cavity shapes: horizontal dissolution cavity (h_1 and h_2 are the depth and height, respectively) (**a**) and vertical dissolution cavity (h_3 and h_4 are the depth and height, respectively) (**b**)

We will also conduct more tests under different test conditions in the future, such as different T stage durations, different upper stress bounds and higher σ_c to judge the stability of rock salt in the triaxial state more accurately.

6.4 Conclusions

Considering the actual stress state of an underground salt cavern used for a CAES facilities, indoor triaxial creep–fatigue tests were designed for rock salt under different confining pressure and stress level conditions, and the main conclusions are as follows:

(1) The addition of confining pressure not only increases the strength of the rock salts but also increases the ultimate deformation of the rock salts. Similar to the stress–strain curves result of uniaxial creep–fatigue tests, the decelerating deformation phase of the triaxial test is also greatly shortened. As the increase in the confining pressure, the rock salts deformation ultimately reveals only one stage.

(2) The occurrence of a high-stress plateaus significantly influences the fatigue residual strain, causing the residual strain in cycle A before the plateaus to be greater than that in the B cycle after the plateaus. As the confining pressure increases, the residual strain difference between cycles A and B shows a decreasing trend.

(3) The presence of fatigue elevates the creep rate of the rock salts. However, the increase in the confining pressure suppresses this effect. With increasing confining pressure, the creep strain of the rock salts during the high-stress plateau indicates an increasing trend at an equal stress ratio.

(4) The brittle–ductile transformation of rock salts under different confining pressure are responsible for the variation in the creep–fatigue mechanical characteristics of the rock salts. When designing the gas pressure level of the CAES salt cavern, the impact of depth on the stress state of the surrounding rocks needs to be taken into account. This is particularly important for vertical dissolution caverns.

References

1. Linjian Ma, Xinyu Liu, Wang Ming-Yang, et al. Experimental investigation of the mechanical properties of rock salt under triaxial cyclic loading[J]. International Journal of Rock Mechanics and Mining Sciences, 2013, 6234–41.
2. John Handin, Hager Jr Rex-V. Experimental deformation of sedimentary rocks under confining pressure: tests at high temperature[J]. AAPG Bulletin, 1958, 42(12): 2892–2934.

3. Ren Song, Yueming Bai, JingPeng Zhang, et al. Experimental investigation of the fatigue properties of salt rock[J]. International Journal of Rock Mechanics and Mining Sciences, 2013, 6468–72.
4. Zongze Li, Zhenyu Yang, JinYang Fan, et al. Fatigue mechanical properties of salt rocks under high stress plateaus: the interaction between creep and fatigue[J]. Rock Mechanics and Rock Engineering, 2022, 55(11): 6627-6642.
5. Jie Chen. Catastrophic Mechanism induced by Damaged Surrounding Rock and Mitigation Principle During bedded salt Cavern Construction Period[D]. Chongqing university, 2012.
6. Mingming He, Ning Li, Caihui Zhu, et al. Experimental investigation and damage modeling of salt rock subjected to fatigue loading[J]. International Journal of Rock Mechanics and Mining Sciences, 2019, 11417–23.
7. Weiguo Liang, Chunhe Yang, Yangsheng Zhao, et al. Experimental investigation of mechanical properties of bedded salt rock[J]. International Journal of Rock Mechanics and Mining Sciences, 2007, 44(3): 400-411
8. G Walton. A new perspective on the brittle–ductile transition of rocks[J]. Rock Mechanics and Rock Engineering, 2021, 54(12): 5993-6006.
9. Shih-Che Yuan, Harrison J-P. An empirical dilatancy index for the dilatant deformation of rock[J]. International Journal of Rock Mechanics and Mining Sciences, 2004, 41(4): 679–686.
10. Aditya Singh, Kumar Chandan, Kannan L-Gopi, et al. Engineering properties of rock salt and simplified closed-form deformation solution for circular opening in rock salt under the true triaxial stress state[J]. Engineering Geology, 2018, 243218–230.
11. Wolfgang-R Wawersik. Determination of steady state creep rates and activation parameters for rock salt[M]. ASTM International, 1985.
12. H Alkan, Cinar Y, Pusch G. Rock salt dilatancy boundary from combined acoustic emission and triaxial compression tests[J]. International Journal of Rock Mechanics and Mining Sciences, 2007, 44(1): 108–119.
13. H Horii, Nemat-Nasser Siavouche. Brittle failure in compression: splitting faulting and brittle-ductile transition[J]. Philosophical Transactions of the Royal Society of London. Series A, Mathematical and Physical Sciences, 1986, 319(1549): 337–374.
14. Mingqing You. Effect of circumferential pressure on the strength of rock specimens and discretization [J]. Chinese Journal of Rock Mechanics and Engineering, 2014, (5): 929-937.
15. Melvin Friedman. Fracture in rock[J]. Reviews of Geophysics, 1975, 13(3): 352-358.
16. E-T Brown, Hoek E. Trends in relationships between measured in-situ stresses and depth[A]// Pergamon, 1978: 211–215.
17. Pierre Bérest, Réveillère Arnaud, Evans David, et al. Review and analysis of historical leakages from storage salt caverns wells[J]. Oil & Gas Science and Technology–Revue d'IFP Energies nouvelles, 2019, 7427.

Open Access This chapter is licensed under the terms of the Creative Commons Attribution-NonCommercial-NoDerivatives 4.0 International License (http://creativecommons.org/licenses/by-nc-nd/4.0/), which permits any noncommercial use, sharing, distribution and reproduction in any medium or format, as long as you give appropriate credit to the original author(s) and the source, provide a link to the Creative Commons license and indicate if you modified the licensed material. You do not have permission under this license to share adapted material derived from this chapter or parts of it.

The images or other third party material in this chapter are included in the chapter's Creative Commons license, unless indicated otherwise in a credit line to the material. If material is not included in the chapter's Creative Commons license and your intended use is not permitted by statutory regulation or exceeds the permitted use, you will need to obtain permission directly from the copyright holder.

Chapter 7
Multi-stage Amplitude Creep–Fatigue Mechanical Characterization of Rock Salt with Acoustic Emission Signal Analysis

During the operation of a salt cavern CAES power plant, the level of gas injection pressure determines the energy storage capacity of the CAES, and the change of gas injection pressure also affects the damage evolution of the surrounding rocks of the reservoir. Acoustic Emission (AE) technique is an effective tool to study the damage evolution patterns of materials such as metals and rocks. When a material is subjected to external loading, the sudden redistribution of stresses (due to microcracking/deformation) converts mechanical energy into acoustic energy, resulting in the generation of elastic waves [1]. This phenomenon is known as acoustic emission and is a concomitant to stress redistribution in internal structures. Rock research on AE first began in 1941, by Obert rock explosion monitoring in mines [2]. However, the AE signal strength of many materials is very weak (a human ear cannot directly hear it), so there is a need to use sensitive electronic instruments to detect, record, analyze AE signals. In 1950, the German scientist Kaiser [3] took the lead in studying the AE characteristics of engineering materials. With the continuous development of electronic technology, the reliability of AE technology is also improving. Scholars around the world have conducted many studies on the mechanical properties of rocks based on AE monitoring of fatigue, compression, tension and creep [4]. Evaluating the structural stability of rock materials by AE signal parameters (counts, energy, peak frequency, duration, etc.) is a widely used method, which is fast and intuitive. And on the fact that by a refine analysis of AE signals, it is possible to determine the location of microcracks, but distinguish different types of damage (size of cracks) [5, 6]. In this Chapter, we will focus on the damage evolution of rock salt under the influence of different confining pressures and stress levels with the help of AE techniques.

7.1 Experimental Methods

7.1.1 Acoustic Emission System

The principle of acoustic emission (AE) monitoring is shown in Fig. 7.1. Elastic waves emitted from an AE source (red microcrack) eventually propagate to reach the surface of the material, causing surface waves that can be detected by AE sensors, which convert the mechanical vibrations of the material into electrical signals that are then amplified, processed, and recorded before being analyzed and extrapolated by researchers, The observed AE signals enables to understand the mechanism by which AE is generated by the material.

The acoustic emission equipment used in our tests is the PCI-2 acoustic emission monitoring system produced by the U.S. Physical Acoustics Company, as shown in Fig. 7.2a. It has a dual-channel detection function, which can simultaneously realize characteristic parameter extraction and waveform processing. The system has 18-bit A/D conversion rate, 1 kHZ–3 mHZ frequency range. It is a new type of AE research tool for scientific research institutes and universities, with the following main features:

(1) Low noise, low power consumption, built-in waveform and HIT processor on the main board, suitable for laboratory research;
(2) Built-in 18-bit A/D converter and processor is more suitable for low threshold (17 dB) setting;
(3) Maximum signal amplitude: 100 dB;
(4) Dynamic range: > 85 dB.
(5) 4 high pass and 6 low pass filters, selectable via software control;

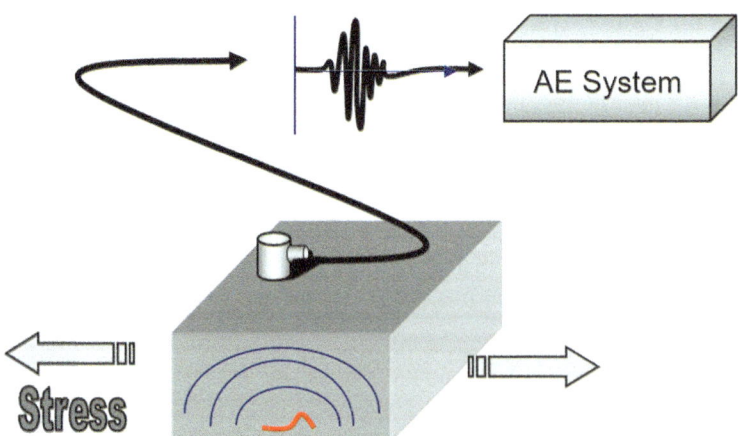

Fig. 7.1 Schematic diagram of the principle of acoustic emission monitoring

7.1 Experimental Methods

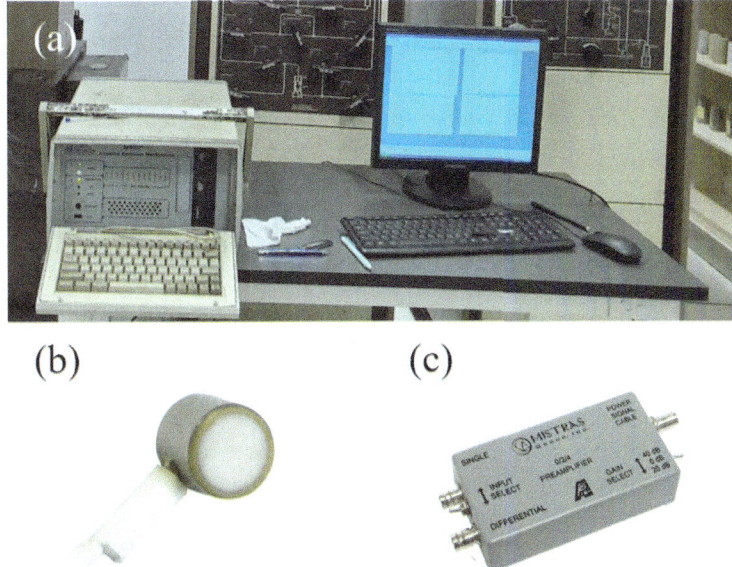

Fig. 7.2 **a** PCI-2 acoustic emission monitoring system, **b** Nano30 miniature AE sensor, and **c** 0/2/4 type advanced universal voltage preamplifier

(6) 40 MHz, 18-bit A/D converter for real-time analysis of samples and higher signal processing accuracy;
(7) The equipment is equipped with AE data flow device, which can continuously store the AE waveforms into the hard disk at a speed of up to 10 M/sec;
(8) The digital signal processor can meet the requirements of high precision and reliability.

The sensor is an important part of the AE detection system, which is an important factor affecting the overall performance of the system. Unreasonable selection of sensors will make the received signal and the actual AE signal have a big difference, which directly affects the realism of the collected data and the data processing results. After comparison and selection, the Nano-30 miniature AE sensor, also from the U.S. Physical Acoustics Corporation, was selected as the AE probe for this test, as shown in Fig. 7.2b. This probe has a resonant response at 300 kHz and a good frequency response in the range of 125–750 kHz, which is suitable for applications in rock mechanics.

Since the voltage signal output from the AE probe is sometimes as low as microvolt orders of magnitude, the signal-to-noise ratio of such a weak signal, after transmission, is bound to be reduced. A preamplifier is set up close to the transducer to boost the signal to a certain level, which is then transmitted via cable to the processing unit of the PCI-2 mainframe. The main technical specifications of the preamplifier are amplification, passband and input noise voltage. The amplifier used in this experiment is the Model 0/2/4 Advanced General Purpose Voltage Preamplifier manufactured by Physical Acoustics, Inc. as shown in Fig. 7.2c. This type of amplifier has switch-selectable gain ranges of 0, 20, and 40 dB. A built-in insertion filter provides the flexibility to optimize transducer selectivity and noise rejection for rock mechanics tests requiring a wide range of gains or frequency bandwidths.

7.1.2 Data Collected by AE

Typical AE signal parameters include amplitude, ring count, duration, energy, threshold voltage value, arrival time, and impact count rate [7], illustrated in Fig. 7.3.

(1) Impact: Any signal that exceeds the threshold and causes a channel to acquire data is called an impact. It reflects the total amount and frequency of AE activity and is often used for AE activity evaluation.

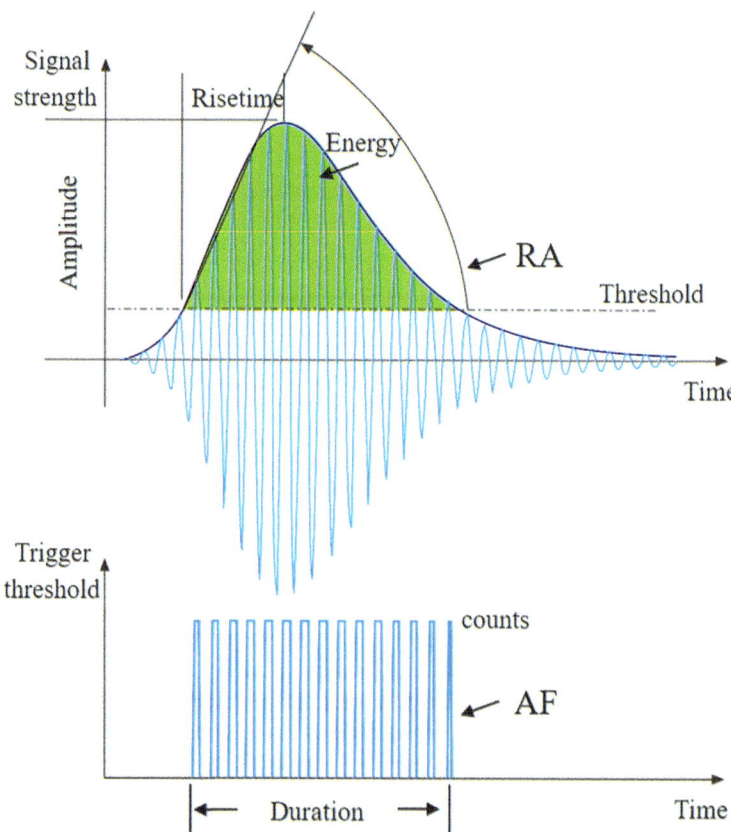

Fig. 7.3 Schematic diagram of typical acoustic emission signal parameters

7.1 Experimental Methods

(2) Count: The same impact is detected by several channels positioned at different location, and the source can therefore be located spatially by analysis of temporal shift.
(3) Ring Count: The number of oscillations above the threshold signal, used for AE activity evaluation.
(4) Energy: The area under the detection envelope of the signal, reflecting the strength of the signal.
(5) Duration: The time interval between the first crossing of the threshold and the final descent of the signal to the threshold.
(6) Rise time: The time from the first crossing of the threshold to the maximum amplitude of the signal. The value of RA is calculated by dividing the rise time by the amplitude of the acoustic emission signal, the unit of measurement is $\mu s * V^{-1}$.
(7) Average Frequency (AF): AF is calculated by dividing the count of acoustic emission events above the threshold by the duration of acoustic emission impacts and is expressed in kHz.

In this Chapter, representative AE parameters (count, energy, RA, AF and so on) will be selected to analyze the damage evolution of rock salt under different loading conditions.

7.1.3 Mechanical Tests Series with AE

The cyclic compressive tests conducted with AE measurements investigated the effect of confining pressure and stress level on the creep–fatigue mechanical properties of rock salt. First, monotonic compression tests were conducted on the rock salt samples at different confining pressures (0 and 3 MPa). Each test was repeated three times, and the average values of the peak stress were computed (see Table 7.1). The results show an average uniaxial compressive strength (UCS) of 30.1 MPa and an average triaxial compressive strength (TCS) under a confining pressure of 3 MPa of 60.2 MPa.

Then, the uniaxial stepwise creep–fatigue (USCF) test and the triaxial stepwise creep–fatigue (TSCF) test were designed as illustrated in Fig. 7.4. These tests were repeated 3 times as well. In the USCF and TSCF tests, the upper and lower stresses were determined based on the compressive strengths of the rock salt under confining pressures of 0 and 3 MPa. The initial upper limit stress was set to 40% peak differential stress of U/TCS defined as stress level (S_l), and the lower limit was set to 3% peak

Table 7.1 Compressive strength of the rock salt under 0 and 3 MPa confining pressure

Confining pressure/MPa	Test no. 1/MPa	Test no. 2/MPa	Test no. 3/MPa	Average/MPa
0	28.9	30.3	31.1	30.1
3	59.5	61.4	59.7	60.2

differential stress of U/TCS; an stepwise increase of 10% per S_l was adopted every 20 cycles. Successive S_l values are given in Table 7.2.

The purpose of such a stepwise creep–fatigue test is to emulate the realistic rheological state of the salt cavern surrounding rocks owing to the prolonged high-pressure condition when gas is injected inside the salt cavern at different pressure levels.

First load to initial S_l is hold for one hour before unloading (T-stage), and then loading applied to the specimen is a repetitive cyclic loading between the low and high stress, with a high-stress plateau duration of 300 s inserted every two stress cycles. There is a rise in the S_l value after every 20 cycles until specimen's failure. Again, the pre-plateau cycle is defined as cycle A, while the post-plateau cycle is referred to as cycle B. The first cycle in every level is neither a pre-plateau nor post-plateau cycle and is called the first (F) cycle. The test results are shown in Table 7.3.

Two Nano 30 AE probes from PAC were installed in symmetrical positions at the middle height of the specimen. A lubricant was applied between the two probes to ensure stable signal acquisition. The AE signals were collected using dual channels throughout the test, from the early occurrence and expansion of microcracks inside the specimen to their coalescence and propagation. The parameters of the AE equipment were set empirically and are given in Table 7.4.

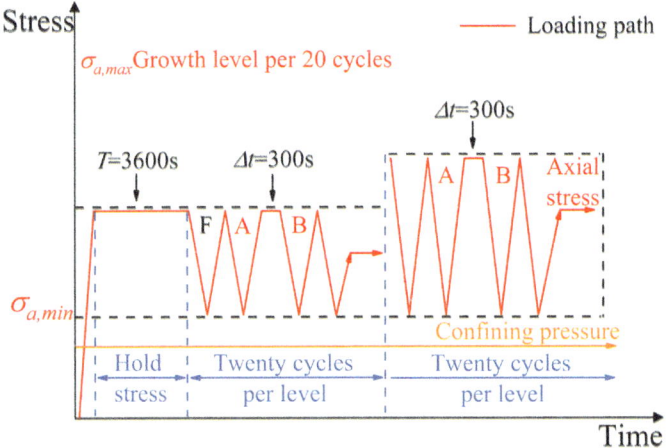

Fig. 7.4 Loading paths for the uniaxial and triaxial stepwise creep–fatigue tests

Table 7.2 Parameters considered in the USCF and TSCF creep–fatigue tests

Test scheme	Upper stress limit/MPa	Lower stress limit/MPa	Plateau times/s
USCF	12, 15, 18, 21, 24, 27, 30	0.9	300
TSCF	25.8, 31.5, 37.2, 42.9, 48.6, 54.3, 60	4.7	

Table 7.3 USCF and TSCF test results

Test number	Confining pressure/MPa	Final stress level	Fatigue life/cycle
U-1	0	5th	88
U-2	0	5th	94
U-3	0	6th	108
T-1	3	7th	126
T-2	3	6th	112
T-3	3	7th	136

Table 7.4 Parameters of the PCI-2 acoustic emission equipment

	Threshold	Peak definition time	Hit definition time	Hit lockout time
Parameters	45 dB	35 μs	150 μs	300 μs

7.2 Results and Analysis of Rock Salt Under Multistage Creep–Fatigue Loading

7.2.1 Stress–Strain Curve of Rock Salt in the U/TSCF Tests

For the following analysis, only representative specimens U-2 and T-3 will be discussed are shown in Fig. 7.5. The final failure of the specimens occurred at 0.8 S_l for the USCF test (24 MPa), and 1.0 S_l for the TSCF test (60 MPa). By observing the two stress–strain curves in the USCF and TSCF tests, we can see that there is a significant increasing in the high-stress plateau strain (creep strain) with increasing S_l. However, for a given S_l, the creep strain decreases with the number of cycles, suggesting a stabilization phenomenon. This is valid for all S_l except the last, during which final failure occurs. Indeed, at 0.8 S_l in the USCF test, the creep strain decreases during the first few cycles and increases again during the later cycles, leading to failure. This final increase is a marker that can help to anticipate final failure. Although this phenomenon is not very obvious, the same pattern can be identified at 1.0 S_l in the TSCF test.

To survey the development on the creep strain rate (v_c) during high-stress plateaus, we counted and calculated the v_c for each high-stress plateau in the USCF and TSCF tests using the same equation as in Sect. 5.2.2. The v_c values for the USCF test are shown in Fig. 7.6.

The black lozenges in the graph represent the creep rate in the first hour (T stage); the creep rate first decreases before gradually stabilizing (creep state), which is due to the low stress level. Subsequently, due to the additional first cyclic load, there is a sudden increase in the creep rate (first red dot), followed by a steady creep rate; then when S_l is increased, the creep rate will jump again and stabilize again etc. The creep rates of the first four stages were calculated, and the creep rate difference between the last two high-stress plateau stages (as shown in the black dotted box in Fig. 7.6)

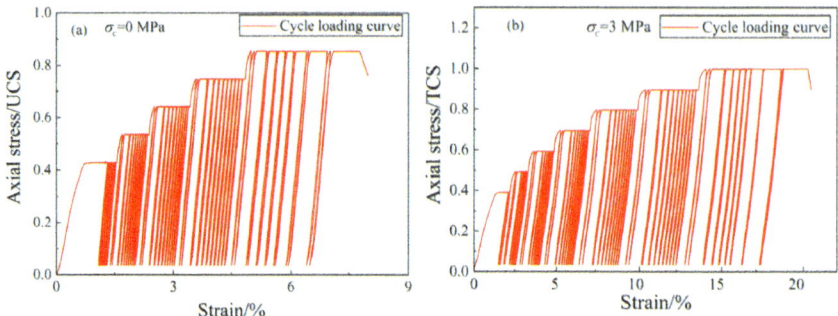

Fig. 7.5 Stress–strain curves from the USCF tests and U-2 (**a**) and TSCF tests T-3 (**b**)

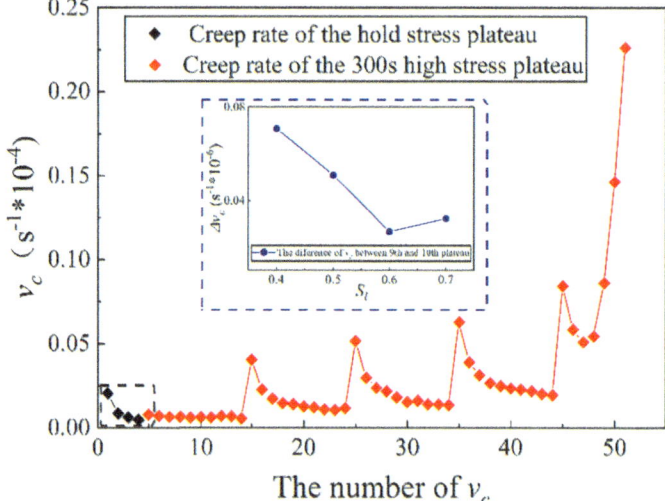

Fig. 7.6 Creep rate during USCF test

for each S_l is shown in the blue dashed box in Fig. 7.6. The difference in the creep rate is very small, with a ratio of < 0.1. This indicates that the rock salt enters the steady-state creep stage again after many cycles. At the same time, it can be seen that with the increase in S_l, the creep rate difference, although fluctuating, shows an overall decreasing trend, indicating that the rock salt specimens are in a compacted and dense state before entering the last Sl. The TSCF and USCF tests present similar results.

7.2.2 Creep Fatigue Residual Strain of Rock Salt in U/TSCF Test

The residual strain ε_r evolutions during the USCF and TSCF tests are shown in Fig. 7.7. The residual strains before (cycle A) and after (cycle B) high stress plateaus in the aforementioned tests are presented for the USCF tests and TSCF tests. Although this phenomenon is not obvious during the first 20 cycles (at 0.4 S_l), for both the USCF and TSCF tests, differences between cycles A and B begin to appear at the second 20 cycles (at 0.5 S_l). This shows that the high-stress plateaus at lower levels still affect the fatigue cycle ε_r of the rock.

In addition, for each S_l, we computed the difference between the $\varepsilon_{r,A}$ and $\varepsilon_{r,B}$ and its average value ($\Delta\bar{\varepsilon}_r$). Because the last few cycles of the final stress level being on the brink of failure, they were not included in the calculation. The results are shown in Fig. 7.8, where the linear regression functions are also proposed.

$$\Delta\bar{\varepsilon}_r = 0.01362 S_l - 0.0036 \quad R^2 = 0.98 \tag{7.1}$$

$$\Delta\bar{\varepsilon}_r = 0.00675 S_l - 0.0013 \quad R^2 = 0.92 \tag{7.2}$$

where Eq. (7.1) is for the USCF test and Eq. (7.2) is for the TSCF test. We find that the $\Delta\bar{\varepsilon}_r$ values in the USCF tests are greater than those in the TSCF tests and that the increasing trend is greater in the USCF test than in the TSCF test.

According to Figs. 7.5 and 7.7, it can be seen that the creep strain and residual strain of the rock salt specimens show an overall increasing trend with increasing stress levels. However, the effect of stress levels on creep strain and residual strain is different. For this purpose, the ratio of creep strain to residual strain was calculated for every S_l (excluding the data at the point of abrupt stress elevation) and defined as φ. The results are shown in Fig. 7.9.

It can be seen that for the first increase in S_l (0.5 level), there is a sudden jump in the ratio (as shown in the black circle), followed by a decrease and then an increase.

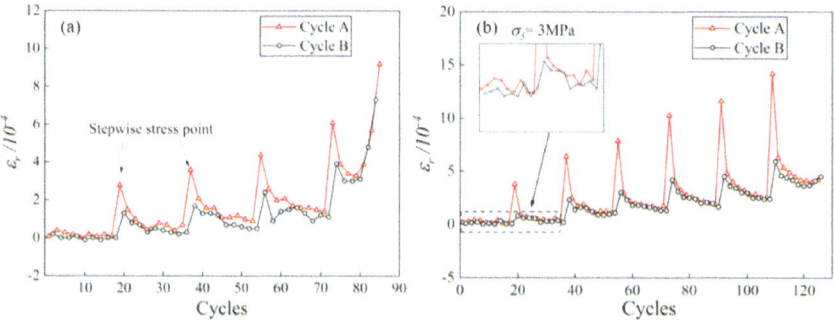

Fig. 7.7 Axial residual strain from each stress cycle during USCF test (**a**) and TSCF test (**b**)

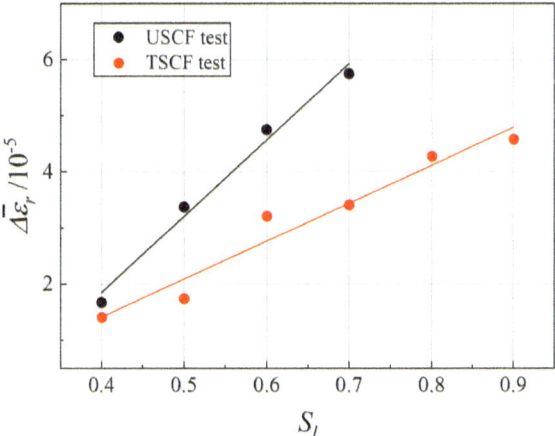

Fig. 7.8 Average residual strain difference between cycles A and B for various increasing stress level S_l, for USCF test and TSCF test

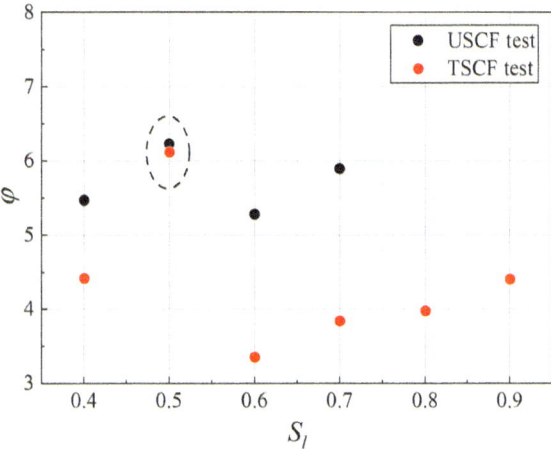

Fig. 7.9 The ratio of creep strain to residual strain φ for every S_l (excluding the data at the point of abrupt stress elevation) for the USCF test and TSCF test

The above phenomenon indicates that the increase in the stress level has more effect on creep than on fatigue. In the second phase, the ratio of creep strain to residual strain peaks. Moreover, it can be observed that the ratio in the TSCF test is consistently lower than that in the USCF test.

7.2.3 AE Counts and Energy During Creep–Fatigue Testing of Rock Salt

The AE counts (N_{AE}) and AE energy (E_{AE}) both can reflect the development of internal microcracks during loading. Figure 7.10 shows the axial strain, the AE counts N_{AE} and the cumulative E_{AE} during the USCF and TSCF tests.

7.2 Results and Analysis of Rock Salt Under Multistage Creep–Fatigue ... 147

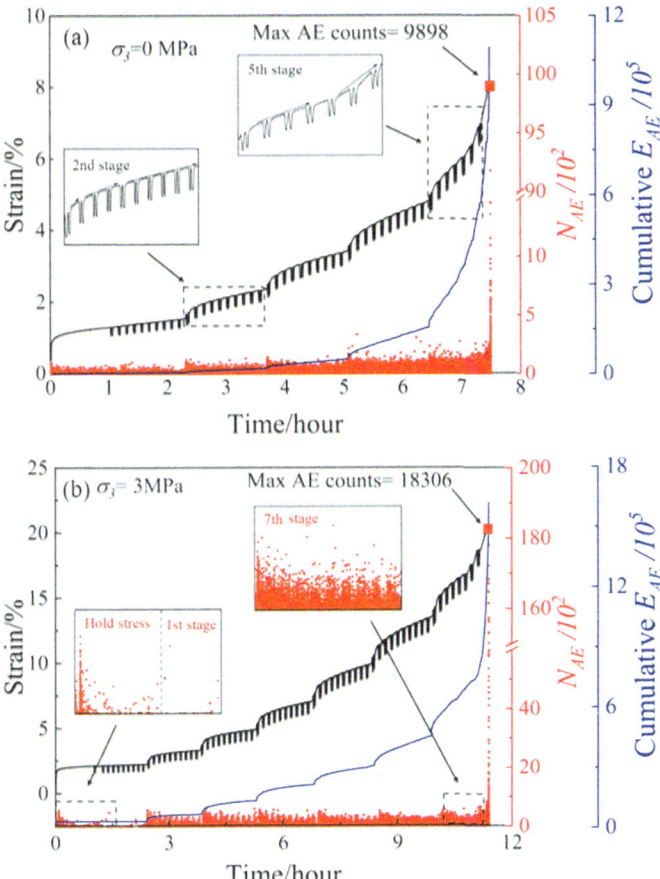

Fig. 7.10 Strain evolution with AE counts and cumulative AE energy for the USCF test (**a**) and TSCF test (**b**)

From the overall N_{AE} evolution and cumulative E_{AE}, the damage of the rock salt is consistent with avalanche dynamics damage theory. The N_{AE} in the final rupture stage is extremely large, arising from the mechanical energy released at the moment of damage. Moreover, the extreme value of N_{AE} for the rock salt in the TSCF test (18,306 counts) is almost twice larger than the N_{AE} in the USCF test (9898 counts), indicating that the confining pressure provides more strain energy that can be released at final failure [8]. In Fig. 7.10a, for the first creep–fatigue cycle of the USCF test, the N_{AE} values are small, and their variation is insignificant compared to the strain. N_{AE} increases significantly at the beginning of $0.5\ S_l$. This is due to the increase in S_l, resulting in the fact that initially closed microcracks subjected to more compaction are reopened and that the slipping crystals are creating new cracks. Write a sentence to describe that for every given stress level, the number of AE events is rather regular, with a small increase at each new cycle beginning. During the last cycle before failure

(0.8 S_l), peaks of N_{AE} occurs several times before final failure. During this stage, larger cracks develop by the convergence of preexisting cracks and grain-piercing cracks from the fatigue process, and the rock salt loses its internal bearing capacity, leading to final failure.

The N_{AE} counts during the TSCF test, plotted in Fig. 7.10b, shows a similar pattern as the USCF test. In particular, at 1,0 S_l, we find that N_{AE} decreases and increases again, following the stress–strain curve in the same cycle (see Sect. 5.2.1). This observation illustrates the reliability of using N_{AE} to characterize rock salt damage.

7.2.4 AE Peak Frequency During the Creep–Fatigue Tests of Rock Salt

The AE signal consists of multiple frequency components, with different waveform frequencies and amplitudes for different AE excitation sources. AEs corresponding to different types and scales of cracks have different spectral characteristics, and the peak frequency (PK) is one of the main indicators used to describe the spectral characteristics [9]. Figure 7.11a and b shows the probability density plots of the frequency distributions for the USCF and TSCF tests, respectively. The PK from AE of rock salt under creep–fatigue cycles are mainly concentrated in three frequency bands: low frequency (0–75 kHz), medium frequency (75–275 kHz), and high frequency (275–400 kHz). When there is no confining pressure, the PK show a rather uniform distribution. The signals firstly appear in the medium-frequency range at low S_l. The low-frequency and high-frequency signals are rare at the beginning of the test and gradually increase for larger values of S_l. The number of medium-frequency signal is almost constant throughout the test, and the probability density reaches its maximum value at final failure.

The frequency distribution in the TSCF test is more distinctly characterized. At the early stage of loading, only sporadic medium-frequency signal distribution present only at the early stage of loading. Thereafter, before the start of the cyclic loading phase, the specimen has a very low probability density, which reflects the compression-density effect of the confining pressure on the rock. Until the end of the test, the AE frequency distribution has a high probability density under the medium frequency condition. The USCF test has a wider distribution at lower frequencies, in the range of 5–75 kHz, while the TSCF test has a wider distribution at higher frequencies, with some distribution above 300 kHz.

It has been pointed out that the magnitude of the AE frequency shows an inverse trend with the crack size [10], i.e., high-frequency signals correspond to small cracks, while low-frequency signals correspond to large cracks. Some studies [11, 12] also have pointed out that tensile damage produces mainly low-frequency AE signals and that shear damage leads to mainly high-frequency AE signals. To observe the

7.2 Results and Analysis of Rock Salt Under Multistage Creep–Fatigue ...

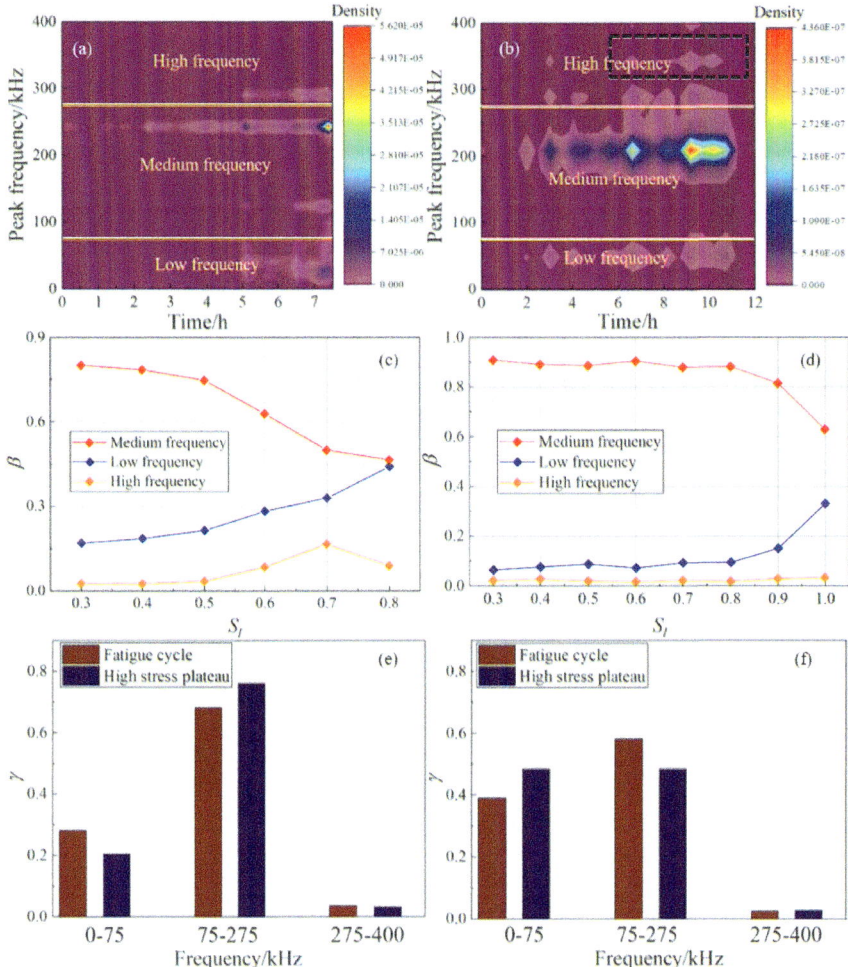

Fig. 7.11 Probability density distribution of peak frequencies and ratio for the three frequency bands for every $S_{l\ levels}$ in the USCF test (**a**, **c**) and TSCF test (**b**, **d**) (*Note* 0.3 S_l is the *T*-stage). Fatigue cycle and high-stress plateau in different frequency bands as a percentage of AE frequency in the USCF test at 0.5 S_l (**e**) and 0.8 S_l (**f**)

distribution of signals at different frequencies during the test phase (from the beginning of the loading until damage occurs), we define β as the ratio of the distribution of AE frequencies in the three frequency bands. This represents the ratio between the number of frequencies occurring in that band and the total number of frequencies. Figure 7.11c shows the β values corresponding to different stress levels during the USCF test. Throughout the test, the medium-frequency signals have the largest proportion, followed by the low-frequency signals and then the high-frequency signals. As S_l increases, the proportion of high-frequency signals gradually decreases (from 0.801 to 0.465), while that of the low-frequency signals gradually increases. In the last stage (damage stage), the proportions of signals in the two frequency bands (high and low) are almost the same. The proportion of high-frequency signals peaks in the stage before failure and then decreases in the last stage. This variation shows that the proportion of large cracks in rock salt specimens gradually increases with increasing S_l during the test and then reaches a maximum in the final damage stage, forming large macrocracks. The development of microcracks reaches a maximum at 0.7 S_l and then decreases again from 0.8 S_l due to the formation of large cracks by coalescence. Meanwhile, the rupture pattern of a specimen in the USCF test can be assessed based on the distribution ratio of the frequency. These results reflect the joint action of tension and shear. Figure 7.11d shows the corresponding values for the TSCF test. The signal proportion of each frequency band in the TSCF test is similar to that in the USCF test in terms of the overall distribution. In this case, the medium-frequency signals have the highest proportion, followed by the low-frequency and then the high-frequency signals. The difference, however, is that the medium-frequency signals in the USCF test have a much larger proportion than that in the TSCF test at 0.3–0.8 S_l, indicating that the size of the cracks created during the triaxial test is smaller than that in the uniaxial test. However, a development pattern similar to that of the USCF test was observed in the final stage, i.e., the proportion of low-frequency signals increases rapidly, demonstrating that the cracks gradually transform into large cracks until failure. The above analysis suggests that the change in frequency distribution may be a precursor indicator of failure in rock salt specimens.

To further analyze the differences in the creep–fatigue response of the rock between fatigue and high-stress plateau cycles under different frequency bands, the frequency proportion in two stages of the USCF test (0.5 and 0.8 S_l) was determined and defined as γ. Figure 7.11e shows the distribution at 0.5 S_l, where a clear difference between the frequency distribution of fatigue and that of creep can be noticed. There is little difference between the two in the high-frequency range, but the proportions for the low- and medium-frequency ranges are different. In the case of low frequency, the proportion of fatigue cycles is greater than that of creep, while for medium frequency, the proportion of creep is greater than that of fatigue. This indicates that the development of very small cracks is more active during the creep stage, while the opening and closing of large cracks are more predominant during the fatigue stage. The percentage at 0.8 S_l shows a different pattern in Fig. 7.11f. There is still not much difference in the case of the high-frequency signals, while for the low-frequency signals, the proportion of creep is larger than that of fatigue. This indicates that the high stress occurring during the creep stage has been transformed into a force that drives crack expansion and formation, which is consistent with previous data.

7.2.5 RA and AF Metrics of Rock Salt During U/TSCF Tests

The risetime/amplitude (RA) and the average frequency (AF) of the AE signal are important parameters that characterize the waveform. These two parameters can be used in the qualitative analysis of the rupture mechanism. Ohno [13] and Nejati et al. [14], investigated the distribution of AF and RA values during the damage of different rocks. Existing studies point out that when the AE signal has a large RA value with a small AF value, it corresponds to a shear wave signal. It can be assumed that the AE event corresponding to this signal is caused by shear damage. Conversely, a small RA value with a large AF value corresponds to tensile damage. The AF versus RA measurements are shown in Fig. 7.12.

Figure 7.13a and b shows the full-field AF and RA distributions for the USCF and TSCF tests, respectively. The USCF test distributions show a highly concentrated feature, with a continuous distribution in the vertical direction, from (0, 0) to (0.25, 695), which indicate tensile damage during failure. In the horizontal direction, the distribution is significantly reduced and characterized by an intermittent distribution, with the presence of very few signals with extremely large RA values and small AF values, corresponding to shear cracking, during the specimen rupture. The distribution of TSCF test signals is similar to that of USCF test signals on the whole, but there are significantly more signals distributed horizontally in the case of the TSCF test. In addition, the extreme value of RA in the TSCF test is larger than that of the USCF test, i.e., there are more shear signals with large RA values and small AF values. To observe the effects of different stress levels on RA-AF signals, the RA-AF distributions at 0.5 and 0.7 S_l in the USCF test and 0.7 S_l in the TSCF test were plotted, as shown in Fig. 7.13c–e, respectively. Comparing the signal distributions under the two stress levels in the USCF test, it can be found that the overall distribution

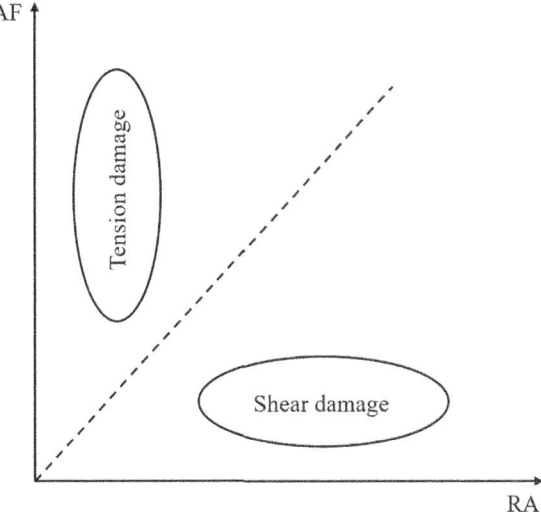

Fig. 7.12 Shear and tensile damage distribution

characteristics of the signals do not change as S_l increases. However, the RA values corresponding to the shear signals show a clear banding feature. The RA values of the second band are distributed among numbers below 6, while the RA values of the fourth band are distributed among numbers above 6. Considering that the increase in stress will obviously promote the development of cracks, it is assumed that the distribution of RA values may reflect the density or scale of crack development. This will be studied in more detail by conducting microscopic observation tests. The signal distribution at 0.7 S_l in the TSCF test is significantly different from that at 0.7 S_l in the USCF test. Signals with high RA and low AF values are significantly more abundant in the TSCF test. The signals show distinct distribution characteristics along the horizontal direction; the RA extreme values are also significantly larger in the TSCF test than in the USCF test. This reflects a more significant shear damage in the TSCF test. Figure 7.13e and f shows the characteristics of RA-AF signal distributions for the fatigue and high-stress plateau cycles in the TSCF test. The signal distribution for the high-stress plateau is more compact, while the signal distribution for fatigue is larger. However, during the last stage, the high-stress plateau produces far more signals than the fatigue cycle. In addition, signals with a large RA value and a small AF value are widely distributed.

7.2.6 Damage Variables Based on AE Counts of Rock Salt During the U/TSCF Tests

Quantifying the correlation between AE and damage from a micromechanical point of view has been a research focus in various countries [15]. As an effective tool to study rock materials, continuum damage mechanics can be used to analyze the damage mechanism of rocks. Kachanov et al. [16, 17] defined the damage variable as:

$$D = \frac{d}{A} \quad (7.3)$$

where D is the damage and refers to the deterioration state of the material, d is the area of the defects appearing on the material and A is the area of the initial undamaged material. The Damage variable D can vary from 0 (no area of defect $d = 0$) to 1 (specimen fully broken so $d = A$). Assuming that the initial state of the rock salt specimen has no damage and that the AE counts are N_t when the whole rock salt specimen is damaged, the AE count rate (δ) per unit area of the load-bearing section can be defined as follows:

$$\delta = \frac{N_t}{A} \quad (7.4)$$

7.2 Results and Analysis of Rock Salt Under Multistage Creep–Fatigue ... 153

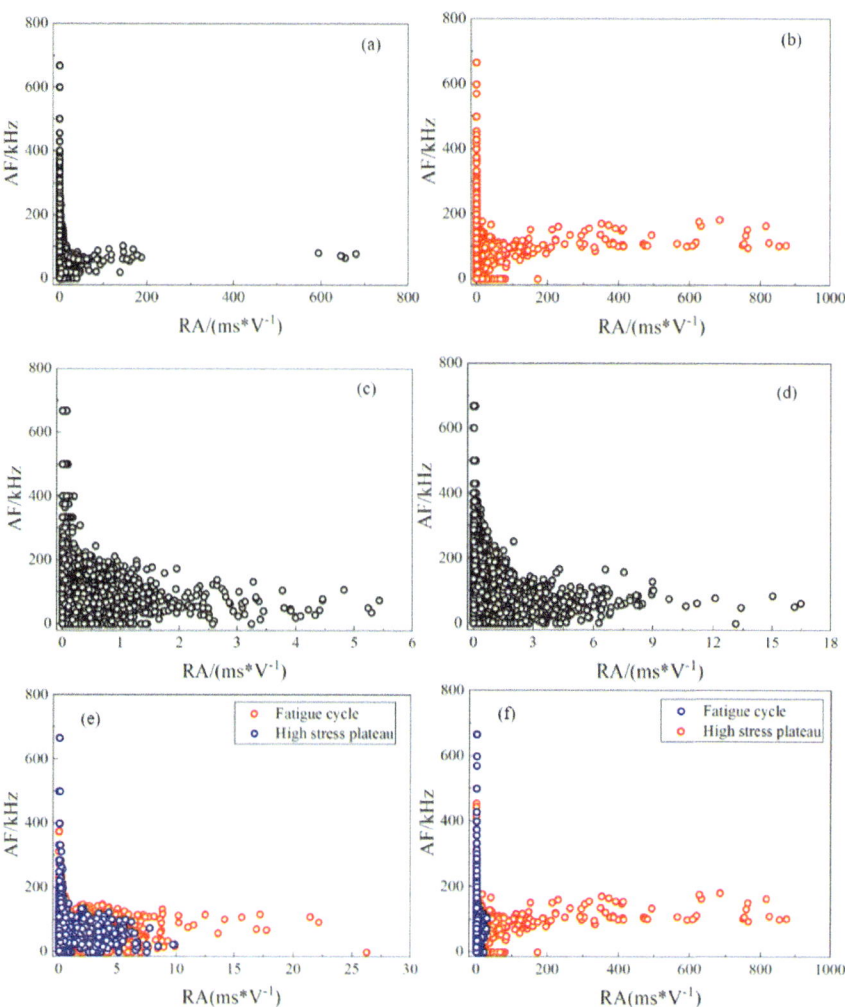

Fig. 7.13 AF versus RA distribution during the USCF test (**a**) and TSCF test (**b**) at 0.5 S_l (**c**) and 0.7 S_l (**d**) in the USCF test and 0.7 S_l (**e**) in the TSCF test. The fatigue cycle and high-stress plateau at 0.7 (**e**) and 1.0 S_l (**f**) in the TSCF test

This equation shows that when the accumulated damage area on the material bearing section is Z_d, the accumulated AE count can be expressed by the following equation:

$$N = \delta * \mathrm{d} = N_t * \frac{d}{A} \tag{7.5}$$

By associating Eqs. (7.4) and (7.5), the relationship between the AE counts N and the damage variable D can be obtained as follows:

$$D = \frac{N}{N_t} \tag{7.6}$$

Equation (7.6) shows that the AE counts can reflect the damage of the material to a certain degree. According to the theory of the Kaiser effect of the appearance of the AE phenomenon, combined with previous studies, the damage variable D_i for the *i-th* cycle is defined as:

$$D_i = \frac{N_i}{N_t} \tag{7.7}$$

The cumulative damage up to the *i-th* cycle can be written as:

$$D_{ti} = \sum_1^i \frac{N_i}{N_t} \tag{7.8}$$

Considering the creep–fatigue test, the test path is slightly different from previous fatigue tests as well as monotonic loading tests. The high-stress plateau also has a large number of AE signals. Therefore, using time as the boundary, the counts in the high-stress plateau stage are counted separately from the N_{AE} in the fatigue phase, and the damage is calculated and then cumulated. The USCF test results are shown in Fig. 7.14.

Figure 7.14a shows the damage distribution characteristics, where D_l is the damage of the first loading as well as the one-hour stress plateau, D_c is the damage of the high-stress plateau, and D_f is the damage of the fatigue cycle, and Fig. 7.14b shows the cumulative damage development. At the beginning of the test, the damage

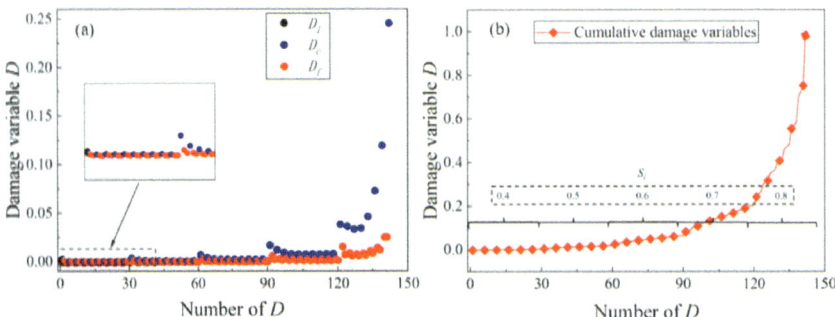

Fig. 7.14 Fatigue cycle and high-stress plateau damage variable development (**a**) and cumulative damage variable (**b**) in the USCF test. Damage includes the initial one-hour high-stress plateau (D_l), fatigue cycles (D_f) and high-stress plateau (D_c)

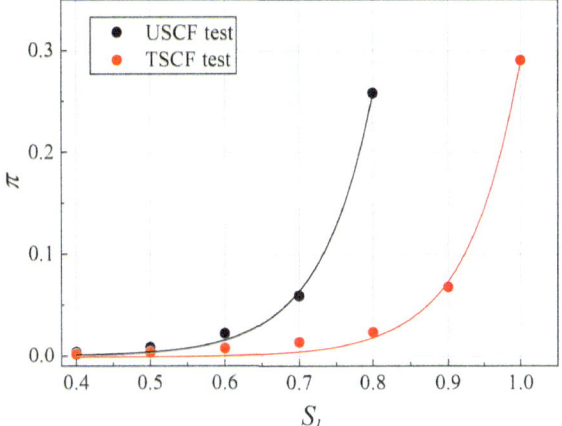

Fig. 7.15 Damage difference π between the creep stage and fatigue cycle for every S_l for the USCF and TSCF tests

in the creep stage was only slightly higher than that in the fatigue cycle, but as the stress increased at 0.7 S_l, the damage in the creep stage was significantly greater than that in the fatigue cycle. In the final stage, the creep stage damage is much larger than that in the fatigue cycle. The cumulative damage shows an almost vertical increase in damage during the final stage. The results for the TSCF test are similar. The test results illustrate that at a lower S_l, the effects in the creep phase and the fatigue cycle are very similar for rock salt; however, as the stress level increases, the creep effect begins to dominate the damage. In addition, the damage difference between the creep stage and the fatigue cycle for each S_l in the USCF and TSCF tests was defined as π.

Figure 7.15 shows the development of π in two different tests. It can be seen that π follows a particular trend. At lower stress levels, π is small, and as S_l increases, the difference increases slowly. However, there is a significant increase occurring in the predamage cycle (0.7 S_l and 0.9 S_l). The curves plotted on experimental data have an exponential form. It can also be noticed that the damage difference in the TSCF test is smaller than that in the USCF test for the same S_l.

7.3 Analysis of Damage Evolution Characterization for Creep–Fatigue Properties of Rock Salts

7.3.1 Discussion on Rock Failure Prediction Based on RA and AF of Rock Salts

During the tests, different crack rupture mechanisms were analyzed qualitatively using two parameters, RA and AF. The signals corresponding to tensile damage and shear damage were different. Therefore, a critical value can be defined to serve as a threshold to differentiate tensile cracking and shear cracking. That is, the slope of the split dashed line k given in Eq. (7.9) was used as an indicator to classify shear and tensile ruptures and to quantitatively describe the distribution of shear and tensile cracks.

$$k = AF/RA \tag{7.9}$$

Many scholars have studied the critical k values of AE to distinguish tension cracking and shear cracking in rocks. Du et al. [18] studied the critical k values of granite, barite, and sandstone based on Brazilian splitting and modified shear tests. The critical k values for granite were also given as 85–95 [19]. However, those authors did not investigate the distribution of k values corresponding to rock salts. In the future, Brazilian splitting testing on rock salt with AE monitoring could be performed to determine the critical k value of rock salt.

Although there are still very few reports on the k values of rock salts, some researchers [20] have proposed another simple method to determine the damage of rock specimens using RA and AF values. Some indoor rock mechanics studies have shown that rocks during the unstable stage tend to have more "shear-type" AE signals with larger RA values and smaller AF values. Therefore, parameter θ, as given in Eq. (7.10), can be used as an indicator for rock damage monitoring.

$$\theta = RA/AF \tag{7.10}$$

An increasing θ value means that the rock is subject to more damage, and extreme values indicate that the rock is in the stage of severe damage.

Figure 7.16a shows the distribution of θ values in the USCF test with time. There is no obvious fluctuation in θ in the early stage, and there is only a slight rise and fall when S_l increases. However, there is an obvious upward trend above 0.7 S_l, indicating that the specimen undergoes more damage. The large number of signals with higher θ values at 0.8 S_l suggests that the rock salt specimen will be destabilized and damaged soon. Figure 7.16b shows the distribution of θ values with time in the TSCF test. Unlike in the USCF test, in the TSCF test, the θ values undergo obvious changes. There is a clear upward movement with increasing S_l. As S_l increases, the maximum value of θ increases, and the number of signals with large θ values also increases, reaching the maximum at failure. Although the phenomenon in the TSCF test is not as clear as that in the USCF test, it can still be seen that for the creep–fatigue test, this parameter can help make some predictions to a certain extent. It can also be seen that θ values change faster than the mechanical response curve. Future studies will therefore focus on the applicability and reliability of the θ value for rock salt.

7.4 Conclusions

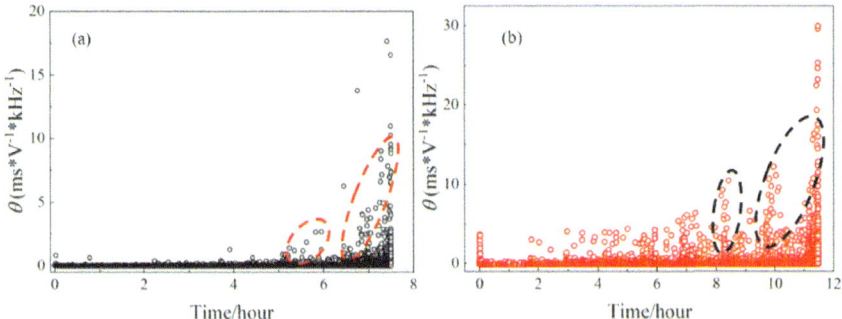

Fig. 7.16 Evolution of θ during USCF test (**a**) and TSCF test (**b**)

7.4 Conclusions

Creep–fatigue mechanical tests were designed reflecting realistic operating gas pressure of a CAES plant (cyclic load with or without plateau, various stress level and various confining pressure levels), and the whole damage process of the rock salt during creep–fatigue testing was monitored using acoustic emissions (AE). The following conclusions were obtained:

(1) The creep strain during the high-stress plateau increases with the stress level. Nevertheless, during every level of cycling, the creep strain showed a decreasing trend with increasing number of cycles, suggesting a stabilization phenomenon. Only during the last stage of cycling, the creep strain show a final increase in strain, leading to specimen failure.

(2) At lower stress levels, the high-stress plateaus have an effect on rock: fatigue begins to manifest with a higher residual strain before the plateau than after the plateau. As the stress level increases, the residual strain values before and after the plateau increase. The presence of confining pressure will reduce this growth trend. The hardening effect during creep, which improves the ability of the specimen to resist deformation, is responsible for this phenomenon in rock salt.

(3) The AE peak frequencies have significantly bar distribution characteristics. The difference in the peak frequency distribution reflects the difference in the failure pattern of the specimens. The specimens in the USCF test predominantly exhibit tensile failure, while specimens in the TSCF test predominantly exhibit shear failure.

(4) The AF/RA magnitude of the AE can be used to determine the crack type. Signals with large RA and small AF values in the triaxial tests are significantly more common than those in the uniaxial tests, reflecting the characteristics of increased shear damage in the TSCF test. Before the failure stage, the signal distribution caused by the high-stress plateau is smaller than that in the fatigue cycle. However, the opposite trend was observed after entering the last stage.

(5) Both the AE parameters and mechanical data show that the plastic strain and damage during the creep stage gradually dominates as the stress level increases. In the final stage, the stress during creep transforms into a force that drives fracture expansion. Therefore, in the design of gas storage, the stress level of the high-stress stabilization stage needs to be considered as an important safety threshold.

References

1. Md-Yeasin Bhuiyan, Lin Bin, Giurgiutiu Victor. Acoustic emission sensor effect and waveform evolution during fatigue crack growth in thin metallic plate[J]. Journal of Intelligent Material Systems and Structures, 2018, 29(7): 1275–1284
2. Leonard Obert. The microseismic method: discovery and early history[A]//1977: 11–12.
3. David-J Holcomb. General theory of the Kaiser effect[A]//Elsevier, 1993: 929–935.
4. Zongze Li, Jinjie Suo, JinYang Fan, et al. Damage evolution of rock salt under multilevel amplitude creep–fatigue loading with acoustic emission monitoring[J]. International Journal of Rock Mechanics and Mining Sciences, 2023, 164: 105346.
5. Patricia Rodríguez, Celestino Tarcisio-B. Application of acoustic emission monitoring and signal analysis to the qualitative and quantitative characterization of the fracturing process in rocks[J]. Engineering Fracture Mechanics, 2019, 21054–69.
6. V-L Shkuratnik, KravChenko O-S, Filimonov Yu-L. Acoustic emission of rock salt at different uniaxial strain rates and under temperature[J]. Journal of Applied Mechanics and Technical Physics, 2020, 61479–485.
7. DEW Stone, Dingwall P-F. Acoustic emission parameters and their interpretation[J]. NDT international, 1977, 10(2): 51–62.
8. Z-H Chen, Tham L-G, Yeung M-R, et al. Confinement effects for damage and failure of brittle rocks[J]. International Journal of Rock Mechanics and Mining Sciences, 2006, 43(8): 1262–1269.
9. Mitiyasu Ohnaka, Mogi Kiyoo. Frequency characteristics of acoustic emission in rocks under uniaxial compression and its relation to the fracturing process to failure[J]. Journal of geophysical research: Solid Earth, 1982, 87(B5): 3873–3884.
10. Y Wang, JQ Han, CH Li. Acoustic emission and CT investigation on fracture evolution of granite containing two flaws subjected to freeze–thaw and cyclic uniaxial increasing-amplitude loading conditions[J]. Construction and Building Materials, 2020, 260119769.
11. Jie Huang, Qianting Hu, ChaoZhong Qin, et al. Pre-peak acoustic emission characteristics of tight sandstone failure under true triaxial stress[J]. Journal of Natural Gas Science and Engineering, 2022, 102104576.
12. LR Li, JH Deng, L Zheng, et al. Dominant frequency characteristics of acoustic emissions in white marble during direct tensile tests[J]. Rock Mechanics and Rock Engineering, 2017, 501337–1346.
13. Kentaro Ohno, Ohtsu Masayasu. Crack classification in concrete based on acoustic emission[J]. Construction and Building Materials, 2010, 24(12): 2339–2346.
14. Hamid-Reza Nejati, Nazerigivi Amin, Sayadi Ahmad-Reza. Physical and mechanical phenomena associated with rock failure in Brazilian Disc Specimens[J]. International Journal of Geological and Environmental Engineering, 2018, 12(1): 35–39.
15. Chengsheng OuYang, Landis Eric, Shah Surendra-P. Damage assessment in concrete using quantitative acoustic emission[J]. Journal of Engineering Mechanics, 1991, 117(11): 2681–2698.
16. Lasar Kachanov. Introduction to continuum damage mechanics[M]. Springer Science & Business Media, 1986.

References

17. W Swindlehurst. Acoustic emission-1 introduction[J]. Non-destructive testing, 1973, 6(3): 152–158.
18. Kun Du, Li Xuefeng, Tao Ming, et al. Experimental study on acoustic emission (AE) characteristics and crack classification during rock fracture in several basic lab tests[J]. International Journal of Rock Mechanics and Mining Sciences, 2020, 133104411.
19. Yixiong Gan, Shunchuan Wu, Yi Ren, et al. Evaluation indexes of granite splitting failure based on RA and AF of AE parameters[J]. Rock and soil mechanics, 2020, 41(07): 2324–2332
20. Haocong Hu, Juanhong Liu, JinAn Wang. Toughness test and acoustic emission characteristics analysis of fiber reinforced concrete[J]. Journal of China Coal Society, 2023, 48(3): 1209–1219

Open Access This chapter is licensed under the terms of the Creative Commons Attribution-NonCommercial-NoDerivatives 4.0 International License (http://creativecommons.org/licenses/by-nc-nd/4.0/), which permits any noncommercial use, sharing, distribution and reproduction in any medium or format, as long as you give appropriate credit to the original author(s) and the source, provide a link to the Creative Commons license and indicate if you modified the licensed material. You do not have permission under this license to share adapted material derived from this chapter or parts of it.

The images or other third party material in this chapter are included in the chapter's Creative Commons license, unless indicated otherwise in a credit line to the material. If material is not included in the chapter's Creative Commons license and your intended use is not permitted by statutory regulation or exceeds the permitted use, you will need to obtain permission directly from the copyright holder.

Chapter 8
Long-Time Creep–Fatigue Mechanical Properties of Rock Salt

In the preceding three chapters, the influences of multiple factors such as high stress intervals (creep) duration, confining pressure, stress levels, etc., on the mechanical characteristics of rock salt creep–fatigue were systematically investigated. The interactions between creep and fatigue were identified. Based on experimental data and a combination of macroscopic and microscopic observations, an analysis was conducted to understand the diverse responses of internal rock salt structures to creep and fatigue, revealing the reasons behind the variations in creep and fatigue damage. The mutual interaction between creep and fatigue leads to a more intricate evolution of damage within the rock salt specimens, a feature also reflected in the results of acoustic emission monitoring.

However, due to the experimental design, the aforementioned tests were conducted within the timeframe of a single working day. Given that the peak shaving frequency of compressed air energy storage power plants often operates on the scale of days [1, 2], conducting long-duration creep–fatigue tests on rock salt becomes crucial for a precise understanding of the damage evolution in salt cavern reservoirs. Hence, in this chapter, with varying peak shaving frequencies and gas storage pressures, three sets of long-duration rock salt creep–fatigue tests were designed. These tests aim to explore the creep–fatigue mechanical characteristics of rock salt on the operational scale of gas storage facilities, contributing to a comprehensive understanding of the salt cavern reservoir's damage evolution.

8.1 Experimental Methods

The loading and unloading path of the test is as follows (see Fig. 8.1): the first stage is the loading stage, the specimen is loaded to the upper limit of stress at a constant stress rate; the second stage is the upper limit of creep stage, keep the upper limit of stress constant and stabilize the pressure to the third stage; the third stage is the

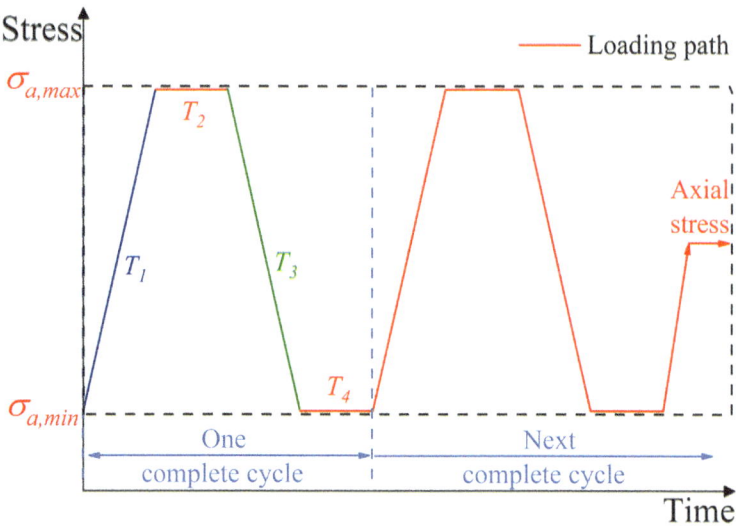

Fig. 8.1 Loading path for long-time creep–fatigue (LTCF) tests on rock salts

unloading stage, with the same constant rate of stress reduced to the lower limit of stress; the fourth stage is the lower limit of creep stage, keep the lower limit of stress constant and stabilize the pressure to the next first stage. The time of each stage is kept the same, and the four stages are a complete cycle, repeating the above four processes until the specimen is destroyed. Comprehensive consideration of salt cavern gas pressure, CAES power plant frequency, regulation cycle, and testing machine working conditions and other factors, the longest test time is set to 8 days. If the specimen is not damaged after 8 days, the test is manually stop.

Uniaxial compressive strength (UCS) is same as previous chapters. The lower stress limit of the test was the same as the previous test, set at 3% of UCS, 1 MPa. The upper stress limit of the test was initially set at 85% of the UCS (test L-1), but it was found that the rock salt specimens were damaged within only 3 days, which could not satisfy the preset target of the test (8 days). Therefore, it was decided that the maximum stress limit was finally decreased to 80% of the UCS for the two other tests L-2 and L-3. The specific loading and unloading parameters for the three tests are shown in Table 8.1. Note that we vary the duration T and the stress level, so the resulting loading rate is different for each test. The detailed description of the stress loading cycle is shown in Eqs. (8.1) and (8.2).

$$T_1 = T_2 = T_3 = T_4 \tag{8.1}$$

$$T_1 + T_2 + T_3 + T_4 = T \tag{8.2}$$

Table 8.1 Loading and unloading parameters in LTCFT

Test number	T/h	Upper stress level/MPa	Loading rate/(kN/s)
L-1	24	24	0.00212
L-2	24	18	0.00157
L-3	8	18	0.00472

The purpose of this setup is to simulate a salt cavern peaking once or three times a day, and the difference in the upper storage pressure limit can be used to observe the effect of changes in the air injection rate as well.

8.2 Results and Analysis of Long-Time Creep–Fatigue Mechanical Properties of Rock Salt

8.2.1 Stress–Strain Curve of Rock Salt Specimen

Figure 8.2 shows the stress–strain curves of rock salt specimens under different maximum cyclic stresses and different loading cycles.

It can be found that the specimen of L-1 test was damaged after entering the 8th cycle, and the specimen completely passed 3 deformation stages of decelerated deformation, stable deformation and accelerated deformation. For the other two tests L-2 and L-3, due to the lower upper stress limit, the specimens were not damaged after 8 days of complete loading, so the stress-strain curves of the specimens only showed the process of 'sparse'–'dense'. i.e., it did not stage III and final failure. Since only L-1 test has reached failure, this test will be taken as the main object of analysis, and the long-time creep–fatigue mechanical properties of rock salt will be investigated by comparing the test results of the two other tests. Observing the stress-strain curves of the three tests, it can be found that the increase of the upper stress limit is decisive for the deformation of rock salt compared with the change in loading rate. In the case of constant upper limit stress, the rock salt specimens with frequent peak shifting produced smaller deformation after the same test time.

Figure 8.3 shows the variation of strain with time for the three tests. The strain curve of the L-1 test shows a rapid rising trend with time, while the second and third tests enter the stage of stable deformation and the strain curves tend to flatten out after a very small number of cycles. Meanwhile, it can be seen that the difference in loading frequency resulted in the strain curves of L-3 tests always lying below the L-2 curves despite the same upper stress limit.

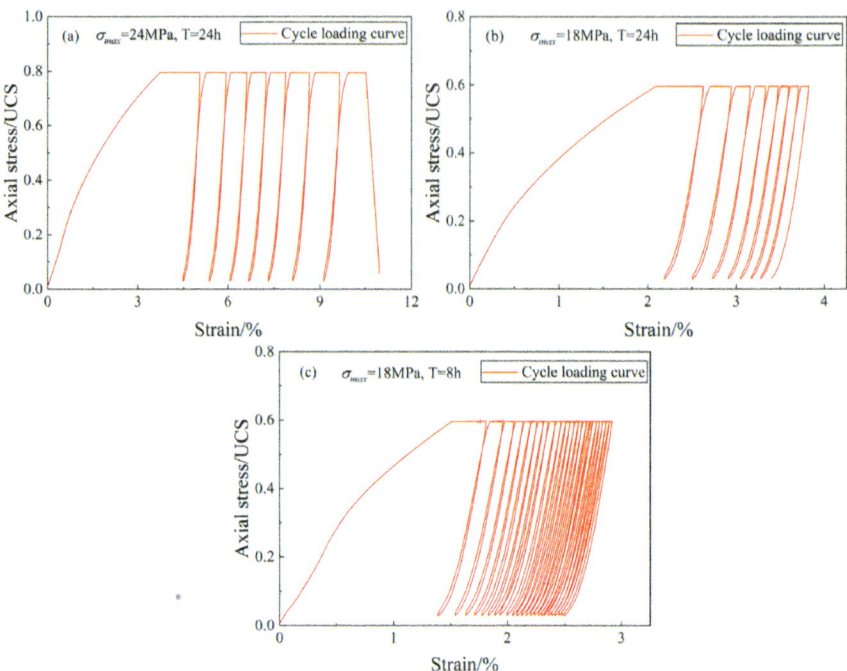

Fig. 8.2 Stress–strain curves for 3 long-time creep–fatigue tests on rock salt

Fig. 8.3 Strain–time curves for three long-time creep–fatigue tests

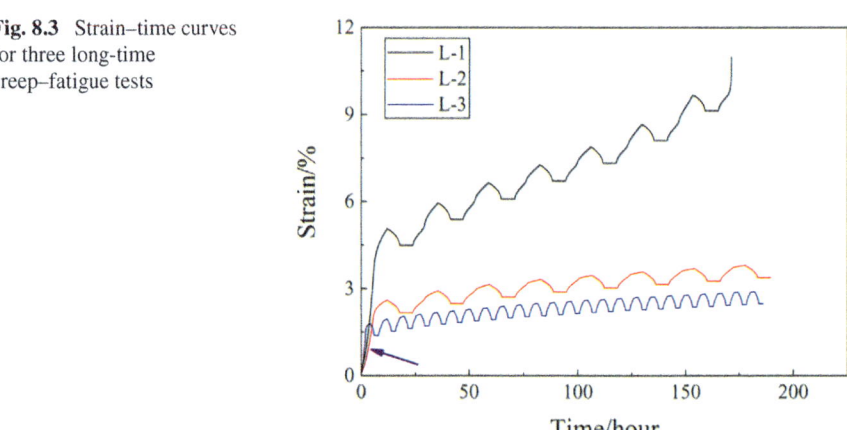

8.2.2 Strain in Rock Salt at Different Loading Stages

In order to clearly show the effect of different loading schemes on the strain development of rock salt specimens, the plastic strains generated at the end of the whole

8.2 Results and Analysis of Long-Time Creep–Fatigue Mechanical ...   165

cycle were calculated by taking the beginning and the end of each complete cycle as the boundaries and the results are shown in Fig. 8.4.

It is noted that due to the large number of cycles in the L-3 test, the method of normalizing the number of cycles (cycle number divided by the total number of cycles of the test) is used to put the vertical coordinates of the three groups of tests on a uniform axis for clear comparison. Anyhow, since L-2 and L-3 tests did not reach final failure, the comparison with test L-1 in terms of horizontal axis remain impossible. It can be seen that as the maximum cyclic stress increases, the plastic strain produced in the specimen at the end of each cycle increases; in the L-1 test, when the specimen undergoes damage, in the cyclic stage before the damage, when the cycle is over, the plastic strain produced increases significantly compared with the previous cycle (see the black dotted box in Fig. 8.4.

Considering that the deformation of the initial cycle is much larger than that of other cycles, the cycle of initial loading to the upper stress limit is named as the first

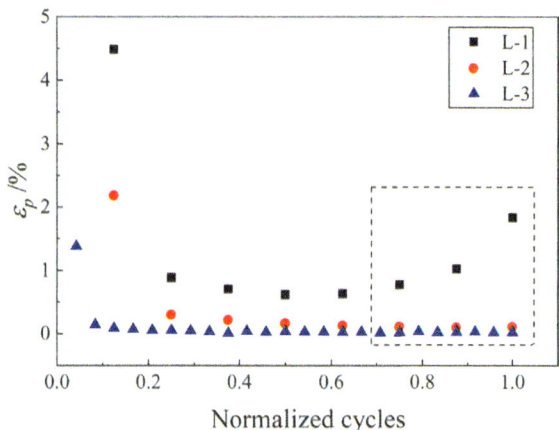

Fig. 8.4 Plastic strain for each complete cycle of the three long-time creep–fatigue tests

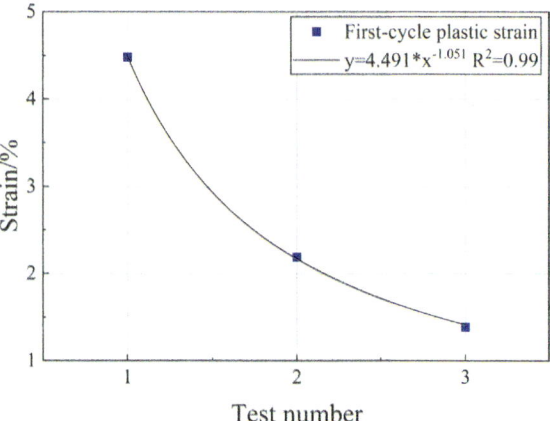

Fig. 8.5 First-cycle strain for three long-time creep–fatigue tests

cycle, and the results are shown in Fig. 8.5. It can be seen that the strains of the three tests show a linear decreasing trend. Since the upper stress limit of the L-1 test is higher than that of the other two tests, the red squares in the figure are the strains of the rock salt specimens loaded to the same upper stress limit, and it can be found that the strains of the rock salt show a decreasing trend with the increase of the loading rate, which is in line with the existing test conclusions, indicating that in the time scale of the salt cavern CAES operation, the rock salt specimen deformation still satisfies that conclusion.

Since the complete cycle in Fig. 8.4 includes both fatigue strains from loading/unloading and creep strains from high/low stress intervals, the strains generated at each stage are calculated separately, and the results are shown in Fig. 8.6.

The most obvious differences between the three sets of tests are in the loading and high stress plateau stage. Firstly, for the complete L-1 test, the deformation in the loading section appeared to be characterized by a distinct second stage and accelerated third deformation stage, where the trend of deformation growth was not obvious for L-2 and L-3 test. In contrast, the development pattern of creep strain is approximately V-shaped, i.e., the strain decreases to a minimum value and does not show a significant stabilizing deformation, which then grows rapidly, and the specimen subsequently undergoes damage. In the L-2 and L-3 tests, which have the same upper stress limit, the loading stage and the high stress plateau stage also appeared to be obviously different, and after the third loading stage, the strain difference between the loading stages of the two groups of tests became very small, which indicates that although the difference in loading rate leads to the difference in the deformation of the rock salt, this effect is weakening as the test proceeds. The speed of the creep strain into the stabilization stage of the specimens in the L-3 test is significantly faster than that of the specimens in the L-2 test, which reflects that, for the same time span, the higher the number of cycles, the faster the creep strain enters the stable deformation phase. That is, the increase in the frequency of peaking contributes to the stabilization of the creep deformation rate of the surrounding rock in the salt cavern. According to this test phenomenon, when the CAES plant needs to change the operating pressure, the peak frequency can be increased appropriately, in order to make the salt carven surrounding rock enter the stable deformation as soon as possible, and to reduce the influence of pressure change on the salt craven surrounding rock. In the unloading stage, the three tests are the same in that the strain remains basically constant with cycling, and the unloading strain of the higher stress L-1 test is obviously larger than that of the other two groups. In the low-stress interval stage, the strain values of the three groups of tests kept fluctuating around the zero.

8.2 Results and Analysis of Long-Time Creep–Fatigue Mechanical ... 167

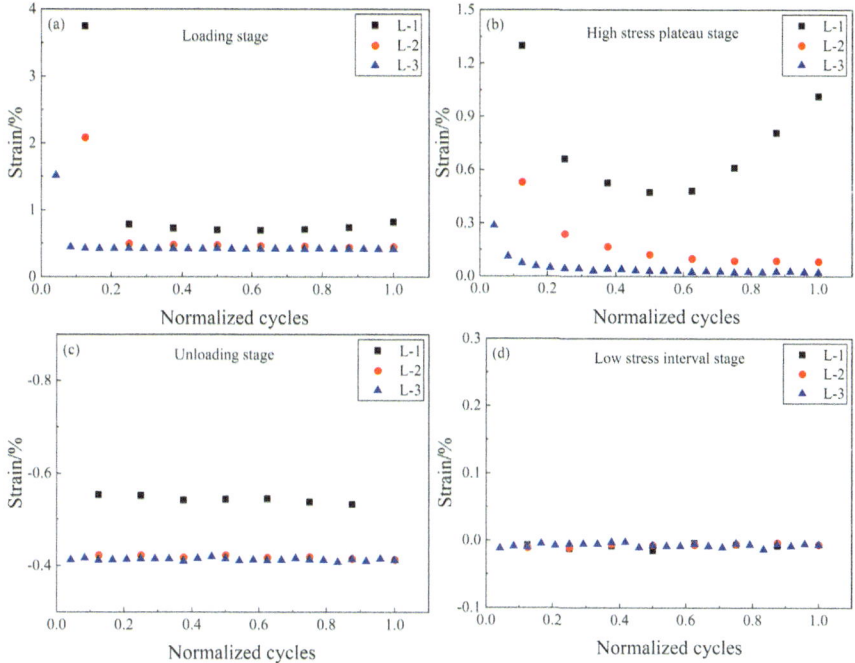

Fig. 8.6 Three long-time creep–fatigue tests: **a** strain during loading, **b** high stress plateau strain, **c** strain during unloading and **d** low stress interval strain

8.2.3 Strain Rate and Elastic Modulus Analysis of Creep–Fatigue Curves

The modulus of elasticity is an important mechanical property parameter of materials. From the macroscopic point of view, the modulus of elasticity is a physical quantity that characterizes the ability of a material to resist deformation under the action of external loading; from the microscopic point of view, the modulus of elasticity is an index that characterizes the bonding strength between ions, atoms, or molecules within a material. Considering that this test is a creep–fatigue test, the loading and unloading also includes a high stress plateau/low stress interval stage, so the elastic modulus is calculated during loading (E_l) and unloading (E_u) phases. The calculations are shown in Eqs. (8.3) and (8.4). The test results are shown in Fig. 8.7.

$$E_l = \frac{\sigma_B - \sigma_A}{\varepsilon_B - \varepsilon_A} \tag{8.3}$$

$$E_u = \frac{\sigma_D - \sigma_C}{\varepsilon_D - \varepsilon_C} \tag{8.4}$$

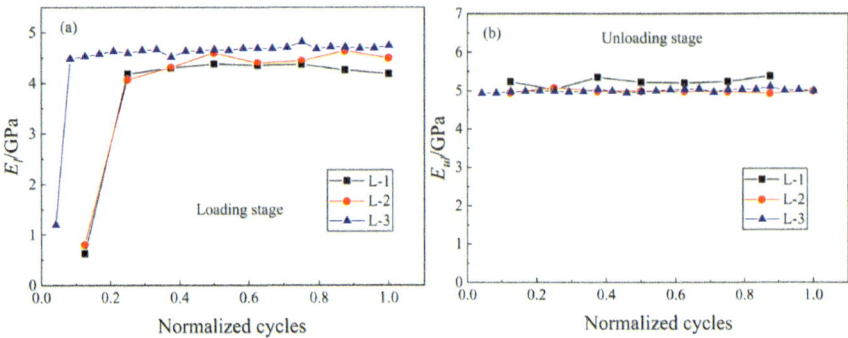

Fig. 8.7 Elastic moduli computed during **a** loading and **b** unloading for three long-time creep–fatigue tests

where E_l is the loading modulus, ε_A and ε_B represent the starting and ending strain values of the approximate linear segment during the loading process, while σ_A and σ_B correspond to the respective stress values. E_u is the unloading modulus, ε_C and ε_D represent the starting and ending strain values of the approximate linear segment during the loading process, while σ_C and σ_D correspond to the respective stress values.

Figure 8.7a shows the change of loading elastic modulus. The first cycle modulus is very low, because the rock salt also produces a large amount of creep deformation. Into the second cycle, the loading modulus of the rock salt tends to reach a stable value between 4 and 4.5 GPa. Here there is a special phenomenon, that is, the modulus of elasticity of the L-1 test is smaller than that of the L-2 and L-3 test. This may be related to the creep inflection point of the specimen. It has been pointed out that the inflection point of accelerated creep of rock salt occurs at the stress level of 0.6–0.7 during the graded creep experiment of rock salt. For this test, it can be assumed that the strain of the rock salt specimen does not increase linearly in the process of stress growth, so it leads to this phenomenon. It can also be found that the loading modulus of the L-1 test shows a decreasing trend in the later part of the test. The unloaded elasticity reflects the ability of the specimen to recover from elastic strain. The larger the unloading modulus is, the less elastic strain is recovered and the more damage is produced in the specimen. Among the three groups of tests, the unloaded modulus of the L-1 test is obviously larger than that of the other two groups (Fig. 8.7b), in addition, the modulus of the L-2 and L-3 tests shows a slight fluctuation with the progress of the test, and the unloaded modulus of the L-1 test shows a rising trend in the last few cycles. This phenomenon reflects the changes in the internal structure of the rock salt specimens.

Furthermore, another phenomenon can be observed in all three test groups, where the unloading modulus is slightly higher than the loading modulus. This is attributed to the fact that the unloading modulus represents the elastic deformation recovery, exhibiting hysteresis effects and being influenced by the presence of plastic deformation during the loading process. To clearly observe the difference between the

8.2 Results and Analysis of Long-Time Creep–Fatigue Mechanical ...

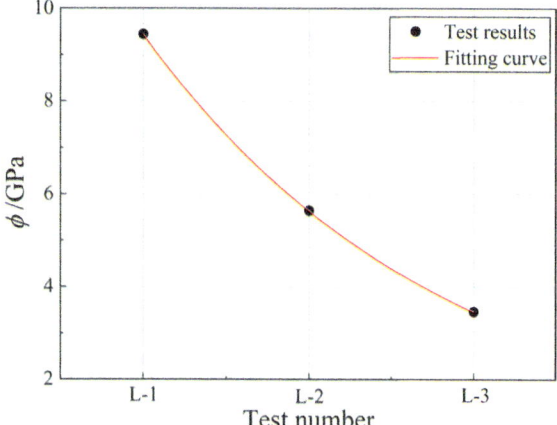

Fig. 8.8 The average of difference between the loaded and unloaded modulus (∅) in the three tests

loading and unloading moduli in each test group, ∅ is defined as the average difference between the loading and unloading moduli (excluding the loading modulus of the first cycle) for each test, as shown in Fig. 8.8.

It can be observed that the stress level has a significant impact on the difference. Due to the fact that rock salt exhibits minimal elastic deformation, higher stress levels result in greater plastic deformation and less elastic recovery, leading to a more noticeable difference between the loading and unloading moduli. Similarly, a faster loading and unloading rate results in less plastic deformation, leading to a smaller difference, indicating a more stable behavior of the rock salt specimen. This phenomenon is closely related to the natural characteristics of the rock.

Rock is a natural engineering material with a large number of internal microscopic defects (microcracks and micropores, etc.), and the existence of these defects is an important factor affecting the change of the modulus of elasticity of the rock under external loading. Generally speaking, the lower the number of microscopic defects inside the rock, the higher the degree of compactness, the stronger the resistance to deformation, and the higher the modulus of elasticity. Therefore, the change rule of elastic modulus of rock salt specimen with the number of load cycles can be explained from the following aspects: (1) when the number of load cycles is less, the original defects inside the rock salt specimen gradually close under the action of cyclic loading, the internal structure gradually tends to be denser, and the resistance to deformation is enhanced to a certain extent, which macroscopically manifests itself in the gradual increase of elastic modulus with the increase of the number of load cycles; (2) when the number of load cycles is increased to a certain degree, the degree of compactness of the rock salt specimen is close to the maximum value under the upper stress, and after that, it basically does not change with the increase of the number of load cycles; at the same time, the rock salt specimen produces relatively few new cracks in this stage, so the modulus of elasticity basically does not change with the increase of the number of load cycles; (3) when it is close to the destruction, the damage inside the rock salt specimen caused by the microcracks and expansion gradually increases, and the ability to resist deformation gradually increases. (4) Near the damage, the damage caused by microcracks sprouting and expanding inside the rock salt specimen gradually increases, and the ability to resist deformation gradually decreases, which is manifested in the gradual decrease of elastic modulus with the increase of the number of load cycles.

In this test, the deformation rate can be differentiated into the loading and unloading stage and the high stress plateau /low stress interval stage, so in order to describe the development of the deformation rate in a unified way, the calculation method is as follows: for all the three groups of tests, the deformation rate is calculated for one hour with the boundary of every hour, and the results of the two groups of tests of the same time cycle, L-1 and L-2, are chosen for graphical analysis in order to make a clear comparison, and the results are shown in Fig. 8.9.

From Fig. 8.9, it can be found that the development of strain rate changes in the deceleration and stable deformation phases of the two sets of tests is similar, and in the first loading section, the strain rate shows a rapid increase with the application of load, which is attributed to the rapid closure of some primary fissures in the interior of the rock salt at the early stage of loading. However, as the loading continues, the internal structure of the specimen is gradually adjusted in place, and the internal resistance to deformation is enhanced, so the phenomenon of slowing down the rate of increase of strain rate occurs. In the L-1 test, the sixth strain rate of the loading section still shows a rising trend, but it is very close to the fifth strain rate, while in the L-2 test, the last strain rate of the loading section is smaller than the fifth strain rate, which indicates that the deformation rate starts to decrease at this time, which leads to this phenomenon, and the hardening of the specimen is related to the beginning of the phenomenon.

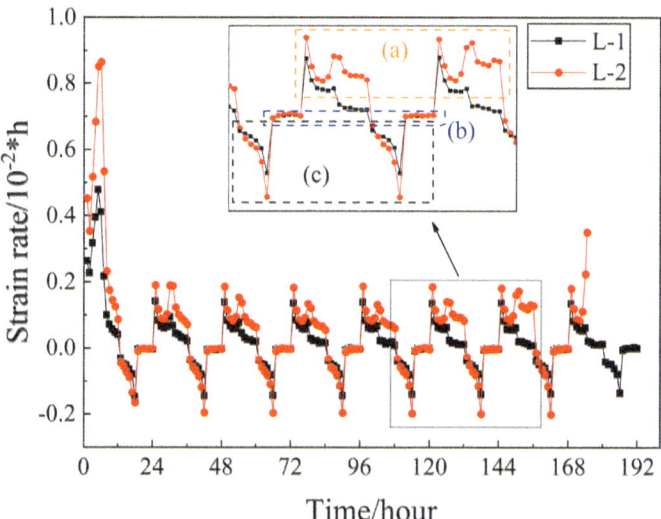

Fig. 8.9 Strain rate versus time for L-1 and L-2 tests

In the high stress plateau, the deformation rate decreases rapidly, but after the specimen passes through the stress adaptation stage, the deformation rate decreasing trend slows down greatly. This pattern is more obvious in the subsequent cycles, indicating that the change of stress state causes a rapid adjustment of the internal structure of the rock salt, but this process will be completed in a relatively short stage of time, and then the rock salt is once again in the process of continuous equilibrium between the internal stress and external load. In the unloading section, the deformation of the rock salt is mainly driven by the internal stress because the external load is continuously removed and the external force is reduced, so the deformation rate is negative, i.e., it is in the state of recovering deformation, and the strain rate is small at this time. For the two sets of tests, the strain rate in the low stress interval section does not change much, basically fluctuating around the 0 point. Due to the large number of data points and the fact that almost all of the rock salt is in the stable deformation stage when it is operated as a air storage. In order to show more clearly, the two groups of tests under the same loading cycle, the different upper limit stress on the stable deformation and accelerated deformation stage of rock salt. The unloading section and low-stress interval of cycles 5 and 6, and the loading section and high-stress plateau of cycles 6 and 7 of the two sets of tests in Fig. 8.9 were taken and enlarged for cross-comparison. As shown in Fig. 8.9a is the strain rate for the loading section and high stress plateau of cycles 6 and 7, (b) is the strain rate for the short low stress interval of cycles 5 and 6, and (c) is the strain rate for the unloading section of cycles 5 and 6.

The orange dashed box in Fig. 8.10a illustrates the variation of deformation rate in the sixth cyclic loading segment for both L-1 and L-2 tests, and it can be noticed that the strain rate increases to a maximum value as the load is applied, and then begins to show a decreasing trend. It is interesting to note that as the stress continues to increase, the fifth strain rate in the loading segment of the L-1 test again shows an increase and the rate of increase is accelerating. Although the increase was small, the last strain rate of the L-2 test still showed an increase. The same phenomenon was observed in the seventh cycle. Similar to what led to the difference in the development of the elastic modulus, this is also related to the inflection point of the stress level for accelerated creep in rock salt. At the fifth strain rate of the L-1 test, where the axial load has been applied for four hours, the stress has grown to more than 60% of the compressive strength of the rock salt, and at the sixth rating it has exceeded 70%, and the stress rating of the L-2 test is also close to 60% of the compressive strength at the sixth strain rate. This is close to the conclusion of existing studies that the inflection point of accelerated creep of rock salt occurs at 60–70% of the compressive strength, after which the creep rate of rock salt increases rapidly.

Meanwhile, it can be found that in the second cycle of L-1 test, the first strain rate of the loading section and the sixth strain rate are almost the same size, but as the test proceeds the first strain rate of the loading section in each cycle shows only a small decreasing trend, but the sixth strain rate shows a significant decrease, and only in the pre-destruction cycle (i.e., the seventh cycle) does it have a tendency to reappear to grow, which shows that in the process of the experiment, the rock salt This indicates that during the experiment, the hardening effect inside the rock salt increases in the

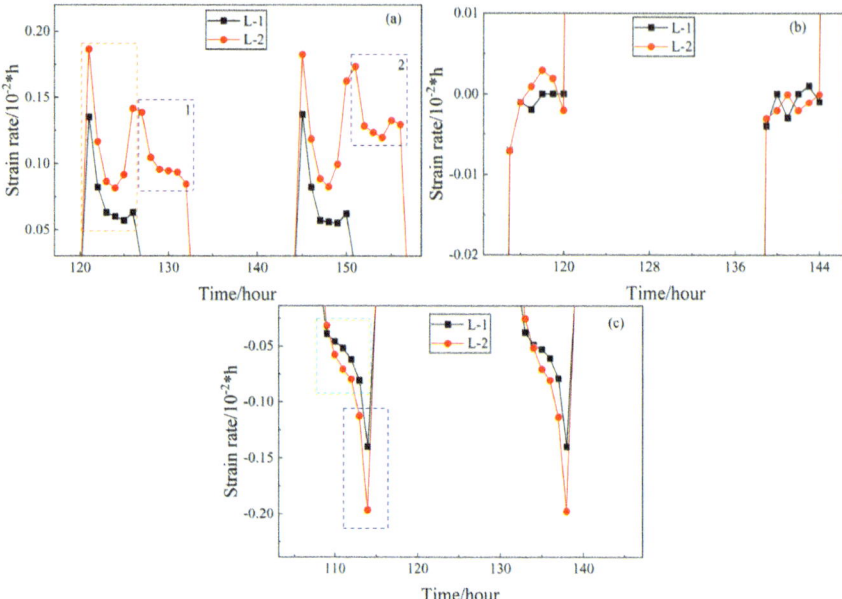

Fig. 8.10 **a** Yellow dashed box of Fig. 8.9, **b** blue dashed box of Fig. 8.9 and **c** black dashed box of Fig. 8.9

first stage, but maintains the equilibrium state in the second stage, and then shows a weakening trend after entering the third stage, at which time the large cracks inside the rock salt are formed, and the cracks penetrate through the surface, and the rock salt gradually loses its bearing capacity and is destroyed. The purple dashed line box is the deformation rate of the high stress plateau, it can be found that for the sixth cycle (including the second to the fifth cycle), the strain rate of the high stress plateau always shows a decreasing trend, but in the seventh cycle, the strain rate of the high stress plateau shows a trend of decreasing and then increasing, at the same time, the strain rate of the other phases has not yet produced obvious changes, which indicates that in the pre-destruction cycle stage, the high stress of the interval has been transformed into the high stress of the pre-destruction phase. high stress in the interval stage has been transformed into a force driving rock destruction. In addition, the creep rate change during the high stress plateau can be used as an indicator to judge the stability state of the surrounding rock in salt cavern. This is an important guide to ensure the stable operation of the salt cavern CAES plant.

Figure 8.10b reflects the development law of deformation rate during the low stress interval of the fifth and sixth cycles, and it can be found that although slightly different, the difference of test conditions has little effect on the strain rate, and the strain rate of the low stress interval of the two groups of tests has a small difference, with a very small value. Figure 8.10c shows the development law of deformation rate in the unloading section of the fifth and sixth cycles, and it can be found that the difference of strain rate in the green dashed box is very small, which indicates that the change of the upper stress limit and the loading rate has a small effect on the initial unloading state of the rock salt, but the unloading rate shows a tendency to increase when the external load is unloaded to a certain stress magnitude and the strain rate reaches a maximum when the external load is unloaded to a minimum value rate reaches the maximum. This is similar for the three sets of test chambers. Moreover, from the second cycle onwards, the development pattern of each cycle is also consistent. The implication of this result is that, as unloading proceeds, the internal structure adjustment of rock salt mainly depends on the differential effects of internal and external stresses, and for the recoverable elastic strain, it basically shows a linear trend according to Hooke's law. When the external force is unloaded to the critical value, the internal stress is greater than the external load, and the deformation rate of the unloaded section of the rock salt grows and grows faster. And it can be noticed that although the L-2 test also showed an increase phenomenon, the final increase value was obviously smaller than that of the L-1 test, which indicates that the difference of the stress state will cause the change of the unloading strain rate. Another feature is that the strain rate in the initial unloading section is smaller compared to the loading section regardless of the test conditions. This provided part of the evidence for the subsequent formulation of the creep–fatigue constitutive equations for rock salts.

8.3 Mechanistic Analysis of Rock Salt Deformation Variations Due to Loading Rates

In this chapter, we discovered that different loading rates lead to variations in the deformation rate and deformation magnitude of rock salt specimens, defining this difference as the rate effect of rock salt. From the above test results, a basic law related to the rate effect can be obtained: as the loading rate decreases, the plastic deformation produced by the rock salt in each cycle increases, and the overall fatigue life decreases. Rock salt is a typical sedimentary evaporite-crystalline rock, mainly composed of NaCl crystals [3]. In the crystal, the plastic deformation of the material is mainly formed through the slip of dislocations, and the larger the number of dislocations and the faster the slip of dislocations, the larger the plastic deformation produced. The movement of dislocations in the crystal is affected by the effective stress and temperature, as shown in Eqs. (8.5) and (8.6), respectively [4, 5].

$$v = B(\sigma *)^{m*} \tag{8.5}$$

$$v = v_c \exp\left(\frac{-\Delta F^{\neq}}{kT}\right) \tag{8.6}$$

$\sigma*$ is the effective stress. $m*$ is a stress sensitivity factor. k is the Boltzmann constant. T is the absolute temperature. $\Delta F^{\neq}$ is standardized activation energy,

Fig. 8.11 Schematic diagram of loading path

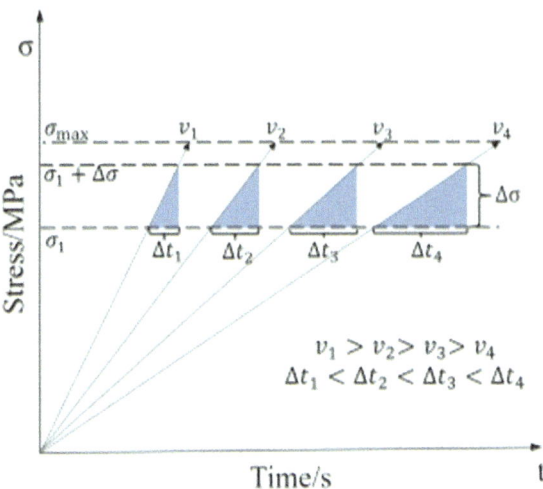

denotes the energy required to make the dislocation fully activated. v_c is the dislocation velocity when the standard activation energy is zero, i.e., the speed of sound. The temperature is approximately constant during the test, and it can be assumed that v is only related to the effective stress.

The loading and unloading paths of the tests in the paper are identical, only the rates are different, i.e., the time required to reach the same stress value is different. The slower the loading and unloading rate, the longer the time required, as shown in Fig. 8.11, the longer the dislocation slip, the longer the slip displacement, i.e., the larger the resulting plastic deformation (residual deformation). It is known from the fourth strength theory of materials, i.e., the distortion energy theory, that the distortion/plastic deformation is the same when the material undergoes damage under the same conditions. The smaller the loading and unloading rate, the greater the plastic deformation produced per cycle, and the smaller the number of cycles the material undergoes at the time of damage, i.e., the smaller the fatigue life.

It should be noted that plastic deformation under the influence of rate effects is not exactly proportional to time or loading and unloading rates. This is due to the fact that during the loading process, the rock salt enters a new stress state from one stress state. Under the new stress state, the rock salt has to gradually proliferate new dislocations to counteract the external force, and the new dislocations are generated to increase the internal force level and reduce the difference between the internal and external forces, so that the dislocations proliferate at a lower rate, and the deformation rate decreases (i.e., deformation stage 1, attenuation deformation stage), until it reaches the balance between the internal and external forces, and enters into the steady-state deformation state. Loading and unloading process stress level actually changes constantly, in a fixed stress level of time is very short, the deformation process is always in the attenuation deformation. As shown in Fig. 8.12, the slow loading rate

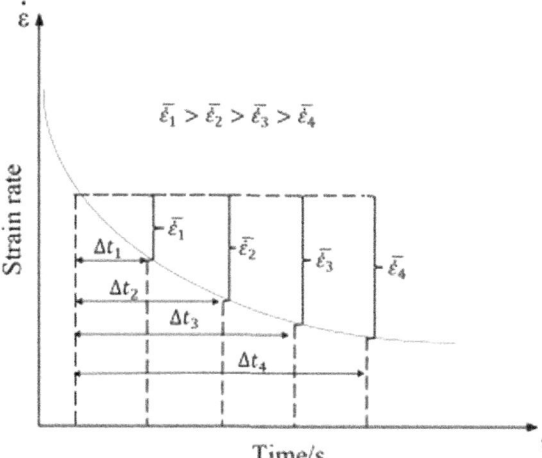

Fig. 8.12 Schematic diagram of strain rate change

corresponds to a lower average deformation rate, but the cumulative deformation is larger.

8.4 Conclusions

The peaking cycles of salt cavern CAES plants are often designed in terms of days, so conducting long-time creep–fatigue tests on rock salt is crucial for accurately evaluating the stability of salt cavern storage. In this chapter, the long-time creep–fatigue tests of rock salt with different loading and unloading cycles and upper stress limits are carried out against the background of the real gas pressure state during the operation of a CAES plant, and the mechanical properties of rock salt creep–fatigue over a long time span are investigated. The specific conclusions are as follows,

(1) In long-time rock salt creep–fatigue tests, the smaller the loading rate, the lesser the deformation generated during the loading phase. The influence of changes in the stress upper limit on rock salt creep is greater than that of the fatigue loading cycle.

(2) The rock salt specimens in the L-1 test underwent a complete three-stage deformation process. Compared to the loading/unloading and low stress interval stages, the variation of strain during the high stress plateau stage is notably significant in the growth phase, exhibiting an overall V-shaped development pattern.

(3) The loading moduli of the three test groups are relatively small during the initial loading. As the tests progress, the loading moduli of L-2 and L-3 tests gradually stabilize, with the loading modulus of the L-3 test showing a decreasing trend. Additionally, both L-2 and L-3 tests have loading moduli greater than L-1, while the unloading moduli exhibit the opposite trend.

(4) The deformation rates during the loading stage and high stress plateau stage of the L-1 test are almost always higher than the deformation rates at corresponding times in the L-2 test. However, in the unloading stage and low stress interval stage, different patterns are observed. The experiments demonstrate that the stress state significantly alters the deformation rate of the rock, but its impact varies depending on different loading/unloading conditions.

References

1. Henrik Lund, Salgi Georges. The role of compressed air energy storage (CAES) in future sustainable energy systems[J]. Energy conversion and management, 2009, 50(5): 1172–1179.
2. David Evans. The Geology, Historical Background, and Developments in CAES[J]. Advances in Energy Storage: Latest Developments from R&D to the Market, 2022: 323–389.
3. Wadi Imseeh, Ran Ma, Timothy Truster, et al. 3D dislocation density–based crystal plasticity model for rock salt under different temperatures and strain rates[J]. Journal of Engineering Mechanics, 2022, 148(3): 04021166.
4. Jingshi Zhao. Fundamentals of Dislocation Theory [M]. National Defense Industry Press, 1989.
5. Deyi Jiang, Zhenyu Yang, Jinyang Fan, et al. Experimental study of load rate effect on fatigue characteristics of salt rock. Rock and Soil Mechanics, 2023, 44(02): 403–414.

Open Access This chapter is licensed under the terms of the Creative Commons Attribution-NonCommercial-NoDerivatives 4.0 International License (http://creativecommons.org/licenses/by-nc-nd/4.0/), which permits any noncommercial use, sharing, distribution and reproduction in any medium or format, as long as you give appropriate credit to the original author(s) and the source, provide a link to the Creative Commons license and indicate if you modified the licensed material. You do not have permission under this license to share adapted material derived from this chapter or parts of it.

The images or other third party material in this chapter are included in the chapter's Creative Commons license, unless indicated otherwise in a credit line to the material. If material is not included in the chapter's Creative Commons license and your intended use is not permitted by statutory regulation or exceeds the permitted use, you will need to obtain permission directly from the copyright holder.

Chapter 9
New Creep–Fatigue Constitutive Modeling of Rock Salt Based on State Variables

Through the studies in the previous chapters, the following main conclusions were obtained: with the increase of stress rate, the plastic deformation of rock salt per cycle decreases, and the deformation of rock salt is distinguished into loading deformation which is affected by the stress rate, and creep deformation which is basically unaffected. In the rock salt creep–fatigue research can be found that there is an interaction between creep and fatigue, fatigue can accelerate the creep deformation, and in turn the hardening effect of the creep stage will reduce the residual strain of the rock salt, the final destruction of the rock salt in the creep–fatigue loading is the result of the dual action of creep and fatigue, when the creep-induced grain perforation cracks and fatigue-induced cracks along the grain boundaries of the fracture or grain perforation fractures will meet and converge, will result in the destruction of the rock salt. Changes in the upper stress limit also have a significant effect on the damage of rock salt specimens. The insight given by the above study is that the special features of the creep–fatigue mechanical ontology model for rock salt and other creep and fatigue ontology models should take into account the effects of creep and fracture extension on fatigue as well as the effects of loading and unloading history on creep. In the following, the salt-rock creep–fatigue constitutive model will be derived and validated for this feature based on some of the laws obtained from the tests.

9.1 Constitutive Modeling of Rock Salt Considering State Variables

9.1.1 Typical Rock Creep Models

(1) Bailey-Norton (Norton) model [1, 2]

Bailey-Norton creep model is one of the time-hardening creep models, which is found to be a better description of the initial and steady state phases of rock creep by comparing with a large amount of experimental data and is expressed in the form of creep strain rate:

$$\dot{\varepsilon} = C_1 \sigma^{C_2} t^{C_3} e^{-C_4/T} \tag{9.1}$$

where $\dot{\varepsilon}$ is the creep strain rate; σ is the equivalent stress; T is the temperature; t is the creep time; C_1, C_2, C_3, C_4 are the parameters of the model respectively.

Since the temperature is constant during the experiment, and hence $e^{-C_4/T}$ is a constant, the above equation can be simplified to the following form:

$$\dot{\varepsilon} = C_1 \sigma^{C_2} t^{C_3} \tag{9.2}$$

The total strain ε can be expressed by the following equation:

$$\varepsilon = \varepsilon_e + \varepsilon_v \tag{9.3}$$

where ε_e is the elastic strain calculated from Hooke's law. ε_v is the viscous strain given by the creep eigenstructure model of Eq. (9.2). The final form of Eq. (9.3) is:

$$\varepsilon = \frac{\sigma}{E} + \frac{C_1 \sigma^{C_2} t^{C_3+1}}{C_3 + 1} \tag{9.4}$$

E is the modulus of elasticity of the rock.

(2) Burgers model [3, 4]

Burgers model can better reflect the transient elastic deformation, viscoelastic deformation, and viscous flow deformation that occurs during the creep process of rock. Burgers model is the classical element model is made of Maxwell and Kelvin bodies in series, and the total strain can be expressed in the following form:

$$\varepsilon = \varepsilon_1(t) + \varepsilon_2(t) \tag{9.5}$$

where ε is the total strain; $\varepsilon_1(t)$ is the strain expressed in Kelvin body; and $\varepsilon_2(t)$ is the strain expressed in Maxwell body. The expansion is shown below:

9.1 Constitutive Modeling of Rock Salt Considering State Variables

$$\varepsilon = \frac{\sigma}{E_1} + \frac{\sigma}{\eta_1} + \frac{\sigma}{E_2}\left[1 - \exp\left(-\frac{E_2}{\eta_2}t\right)\right] \quad (9.6)$$

η_1, η_2 are the coefficients of viscosity of the rock, respectively.

(3) Nishihara model [5, 6]

The Nishihara model is commonly used to describe the deformation of soft rocks, and is also an elemental model with a structure formed by Hooker, Kelvin, and Bingham bodies in series, and the total strain can be expressed in the following form:

$$\varepsilon = \varepsilon_e + \varepsilon_{ve} + \varepsilon_{vp} \quad (9.7)$$

where ε_e, ε_{ve}, ε_{vp} denote the strains of the Hooker body, viscoelastic body, and viscoplastic body, respectively. The expansion is shown below:

$$\varepsilon = \frac{\sigma}{E} + \frac{\sigma}{E_3}\left[1 - \exp\left(-\frac{E_1}{\eta_1}t\right)\right] \quad \sigma > \sigma_s \quad (9.8)$$

$$\varepsilon = \frac{\sigma}{E} + \frac{\sigma}{E_3}\left[1 - \exp\left(-\frac{E_1}{\eta_1}t\right)\right] + \frac{\sigma - \sigma_s}{\eta_2}t \quad \sigma < \sigma_s \quad (9.9)$$

σ_s is the yield stress.

Rock salt has good creep deformation characteristics, and can maintain high strength within a certain range of deformation, which is conducive to the stable operation of the gas storage reservoir [7]. The typical creep curve is divided into three parts: transient creep, steady state creep and accelerated creep, and the operation time of salt cavern storage reservoirs is generally up to 30–50 years [8]. Compared with the entire service life, the transient creep time of rock salt is very short, therefore, in the actual engineering generally only consider the damage in the steady state creep stage and the accelerated creep stage, when the damage or creep develops to a certain stage, the creep enters the accelerated stage [9, 10]. Compared with other models Bailey-Norton model has been used a lot in describing the steady state creep rate of rock salt due to fewer parameters, which are easy to be obtained from experiments, and the model has been widely used in many numerical calculations and engineering evaluations. The present salt-rock creep–fatigue model is derived based on the Norton model, which also creates conditions for future numerical embedding.

The Norton model [11] is mainly used to describe the relationship between stress and strain in the steady-state creep phase of rock-like materials:

$$\dot{\varepsilon} = a\sigma^n \quad (9.10)$$

where $\dot{\varepsilon}$ denotes the strain rate, i.e., the first-order derivative of strain with respect to time, σ denotes the stress, a denotes the material coefficient, and n is the stress coefficient. Although the Norton model is able to make good predictions for the phase characteristics of deformation, there are several problems such as:

(1) The model mainly focuses on the first and second stages of deformation, i.e., the deceleration deformation stage and the stabilization deformation stage. The prediction error is large for the last stage of deformation (third stage), i.e., the accelerated deformation stage before final failure.
(2) The model mainly considers the loading process, and does not fully consider for the unloading process. The result is that the loading and unloading produce exactly the same viscoplastic deformation, which is not consistent with the experimental observations.
(3) The model does not consider the influence of historical loading and unloading stress paths on viscoplastic behavior, i.e., historical cumulative damage or cumulative hardening/softening characteristics are not taken into account in the model.

For these reasons, a new creep intrinsic model is developed to describe the creep–fatigue stress–strain relationship of rock salts affected by unloading stress paths, taking into account the physical and mechanical properties of rock salts.

9.1.2 Definition of Plasticity Factor (State Variable)

Component models tend to have a more explicit physical meaning [12, 13]. A general viscoelastic-plastic model, as shown in Fig. 9.1, consists of a spring component, a friction component, and a damping component, where the spring component (E) characterizes the stress–strain relationship of elasticity, the friction component (σ_{s1}) characterizing loading plasticity, the friction component (σ_{s0}) characterizes the plastic behavior of the material, and the damping component (η_2) usually characterizes the viscous nature of the strain rate proportional to the rate.

Damping component (η_2) represents the variable-rate creep process of the rock salt, specifically the first stage of deformation, where the specimen is undergoing continuous hardening, resulting in a gradual decrease in deformation capacity, indicating the decelerated deformation stage. Once the specimen enters the steady-state stage, the decelerated creep disappears, indicating a correlation between state changes

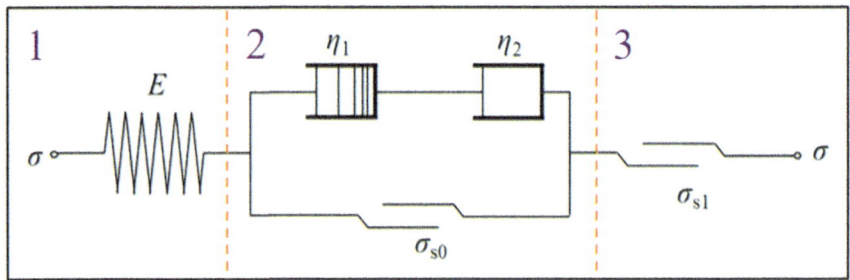

Fig. 9.1 Creep–fatigue rheological model

9.1 Constitutive Modeling of Rock Salt Considering State Variables

and time. When stress undergoes further changes, the material requires a certain time to enter a new steady-state stage, suggesting a correlation with stress as well. Here, the concept of state variable is introduced.

During loading and deformation of rocks occurs a process of continuous loss of elastic deformation capacity, i.e., the process of increasing plastic deformation, so a plasticity factor is the state variable in the process of loading of rocks, and this state variable characterizes the elastic recovery capacity of rocks, which can be understood as the unloading elastic modulus of standard rock specimens.

(1) Effect of loading

Rock salt is a sedimentary rock formed by evaporation as a polycrystal of sodium chloride, and its deformability is related to the density of internal dislocations. Dislocations determine the magnitude of creep deformation; dislocation density is related to the magnitude of the internal force in the deformation zone (i.e., the region where dislocation distortion is produced) within the rock salt [14, 15]. When the external force is larger than the internal force in the deformation zone, the crystal proliferates and produces more dislocations, while the internal force is elevated and the deformation rate gradually slows down, i.e., the deceleration creep phase. When the external force is balanced with the internal force, the dislocation proliferation and slip tend to stabilize, and the deformation rate reaches a stable stage, i.e., the steady-state stage. The state variable σ^* is defined here to characterize the degree of plastic deformation related to the dislocation density within the rock salt. The state variable can be written in the following functional form

$$\sigma^* = \frac{\sigma}{1 + c\sigma^m t^k} \tag{9.11}$$

where c, m, and k are material coefficients; t is the testing time. The above formula can be transformed into the following formula:

$$\frac{\sigma^*}{\sigma} = \frac{1}{1 + c\sigma^m t^k} \tag{9.12}$$

Observing Eq. (9.12), it can be found that if the stress is constant, the state variable will approach the external force σ toward infinite time. When the stress changes (here, only the loading process is considered for the time being), the equation is no longer applicable. The effect of the accumulated plastic deformation needs to be considered. Assuming that the state variable before the new stress action is σ_0^*, then the following equation can be obtained:

$$\sigma_0^* = \frac{\sigma}{1 + c\sigma^m (t_0)^k} \tag{9.13}$$

$$1 + c\sigma^m (t_0)^k = \frac{\sigma}{\sigma_0^*} \tag{9.14}$$

$$c\sigma^m(t_0)^k = \frac{\sigma - \sigma_0^*}{\sigma_0^*} \tag{9.15}$$

$$(t_0)^k = \frac{\sigma - \sigma_0^*}{c\sigma^m \sigma_0^*} \tag{9.16}$$

$$t_0 = \left(\frac{\sigma - \sigma_0^*}{c\sigma^m \sigma_0^*}\right)^{1/k} \tag{9.17}$$

where t_0 is the time corresponding to the state variable σ_0^*. Then after a period of time the new state deformation can be expressed as:

$$\sigma^* = \frac{\sigma}{1 + c\sigma^m(t + t_0)^k} + \sigma_0^* \tag{9.18}$$

According to the above study, the plasticity factor is a state variable characterized by stress and time, and the effect of creep–fatigue interaction on the deformation of rock salt is achieved by changing the value of the state variable, and the intrinsic model established on this basis considers the influence of the loading and unloading path and time on the deformation.

The deformation in the first stage, i.e., the deceleration deformation stage, is mainly influenced by the rapidly generated dislocations, i.e., the plasticity factor, which can be expressed by drawing on the steady-state creep stress–strain relationship of Norton's model as

$$\varepsilon = b\frac{\sigma^*}{\sigma}\sigma^n \tag{9.19}$$

where b is the material coefficient. In the second stage of stable deformation, the creep rate is almost constant and the Norton model can be used directly, the strain given in Eq. (9.10) at this time can be expressed as

$$\varepsilon = \dot{\varepsilon}t = at\sigma^n \tag{9.20}$$

The joint Eq. (9.19) and (9.20) give both the stress–strain relationship between the deceleration deformation phase and the stabilization deformation phase

$$\varepsilon = \sigma^n\left(at + \frac{\sigma^*}{\sigma}b\right) \tag{9.21}$$

The above equation can be rewritten as

$$\varepsilon = \sigma^n\left(at + \frac{b}{1 + c\sigma^m t^k}\right) \tag{9.22}$$

9.1 Constitutive Modeling of Rock Salt Considering State Variables

The relationship between the strain rate and the stress and state variables can be obtained by finding the first order derivative of time for Eq. (9.22)

$$\frac{\partial \varepsilon}{\partial t} = \dot{\varepsilon} = \sigma^n \left\{ a - \frac{bck\sigma^m(t+t_0)^{k-1}}{\left[1+c\sigma^m(t+t_0)^k\right]^2} \right\} \tag{9.23}$$

The relationship between the state variable rate and stress can be obtained by finding the first order derivative of time for Eq. (9.18)

$$\frac{\partial \sigma^*}{\partial t} = \dot{\sigma}^* = -\frac{ck\sigma^{m+1}(t+t_0)^{k-1}}{\left[1+c\sigma^m(t+t_0)^k\right]^2} \tag{9.24}$$

From the above Eqs. (9.23) and (9.24), it can be seen that the strain rate changes accordingly when the stress changes. During loading, if the stress loading rate is v, the stress at time t can be expressed as:

$$\sigma = vt + \sigma_0 \tag{9.25}$$

where σ_0 is the initial stress. Then the state variables can be obtained by integrating the following equation:

$$\sigma^* = ck \int \frac{(\sigma_0 + vt)^{m+1}(t+t_0)^{k-1}}{\left[1+c(\sigma_0+vt)^m(t+t_0)^k\right]^2} dt + \sigma_0^* \tag{9.26}$$

Then the strain during loading can be obtained by integrating the following equation

$$\varepsilon = \int (\sigma_0 + vt)^n \left\{ a - \frac{bck(\sigma_0+vt)^m(t+t_0)^{k-1}}{\left[1+c(\sigma_0+vt)^m(t+t_0)^k\right]^2} \right\} dt + \varepsilon_0 \tag{9.27}$$

where ε_0 is the initial strain. The above equation can be simplified to obtain

$$\varepsilon = \int (\sigma_0 + vt)^n \left(a + \frac{b\sigma^*}{\sigma_0 + vt} \right) dt + \varepsilon_0 \tag{9.28}$$

(2) Effect of unloading stage

In the unloading state, there may be a situation where the external force is less than the state variable, when the material is in the "over-hard" state internally, the dislocation occurs in reverse motion, and the local dislocation density gradually decreases, which is the state variable of:

$$\sigma^* = \frac{\sigma}{1 - c\sigma^m(t+t_0)^k} \tag{9.29}$$

Considering that when unloading, the external force is withdrawn, the internal structure of the rock salt is adjusted completely by internal force, and the adjustment rate is slow, introduce the variable speed creep unloading factor (state variable unloading factor) U_1, find the first order derivative of time, and get the new state variable rate as

$$\sigma^* = U_1 \frac{ck\sigma^{m+1}(t+t_0)^{k-1}}{\left[1 - c\sigma^m(t+t_0)^k\right]^2} \quad (9.30)$$

$$t_0 = \left(\frac{\sigma_0^* - \sigma}{c\sigma^m \sigma_0^*}\right)^{1/k} \quad (9.31)$$

The deformation rate of stable creep is also affected by the introduction of a stable creep unloading factor U_2, and the strain integral form can be rewritten as

$$\varepsilon = \int (\sigma_0 + vt)^n \left\{ aU_2 + \frac{U_1 bck(\sigma_0 + vt)^m \left(t + \sigma_0^{-*}\right)^{k-1}}{\left[1 - c(\sigma_0 + vt)^m \left(t + \sigma_0^{-*}\right)^k\right]^2} \right\} dt + \varepsilon_0 \quad (9.32)$$

(3) **Stress–strain relationships in the accelerated creep phase**

Before considering the third stage, i.e., the accelerated creep stage, we need to first explain the damage of the rock. The concept of damage to materials was originally proposed by Kachanov in 1958 and refers to the structural changes in materials under the action of temperature, load, environment, time and other factors that cause defects, resulting in the deterioration of the mechanical properties of the material and eventually the formation of macroscopic material damage [16]. From the point of view of fine physics, damage is the result of the formation and development of defects such as dislocations, slips, microporosity and cracks experienced by the material [17, 18]. Macroscopically, the loss of load-bearing capacity of a part of the material is considered damage, as defined here

$$D = \frac{d}{A} \quad (9.33)$$

D is the material damage, d is the area without load-bearing capacity material, that is, the damage area, A is the initial area of the material. Assuming that the initial porosity of the material is f, at this time the material, in addition to the pores, are the initial loadable area, it can be considered that the damage is

$$D = \frac{f}{A} \quad (9.34)$$

Due to the special geological structure of rock salts, their microscopic spatial structure is extremely dense and the porosity is almost zero, so the initial damage

9.1 Constitutive Modeling of Rock Salt Considering State Variables

can be considered as zero. The main mechanism of crystal plastic deformation is dislocation slip. During dislocation slip, the corresponding material part does not produce significant damage loss of energy. Here the formation of cracks, holes and other defects that make the material lose its load-bearing capacity is mainly considered as damage, and dislocations are considered as the development of damage, not actual damage. And crack formation is mainly generated by the accumulation of dislocations, a large number of dislocations plugging the stress field generated and the superposition of the external stress field formed by the strength factor to reach the fracture toughness, that is, the generation of fractures (or called fracture nucleation) [19, 20]. The fracture nucleation has not been observed directly from rock experiments and is tentatively considered to be a rapid and instantaneous process. It is assumed that the initial cleavage shaped nucleus has a length of d_0 and is formed instantaneously. The moment of its formation is related to the externally applied stress and accumulated plastic deformation. At this point the damage can be defined as:

$$D = \frac{d_0}{A} \tag{9.35}$$

Assuming that the material produces a plastic deformation rate of $\dot{\varepsilon}$, the rate of change of damage area $\dot{d}$ can be expressed as:

$$\dot{d} = A\dot{D} = \mu_d \dot{\varepsilon} \tag{9.36}$$

where $\dot{D}$ is the damage rate, μ_d is the crack expansion factor, i.e. the actual crack coefficient due to plastic deformation, with the same magnitude as the area. Then the crack extension area d can be calculated by the following formula:

$$d = \mu_d \int \dot{\varepsilon} dt + d_0 \tag{9.37}$$

After the damage, the bearing area of the material decreases and the corresponding stress changes, i.e., the effective stress σ_{eff}^n can be found by the following equation:

$$\sigma_{eff}^n (1 - D) + \lambda D \sigma = \sigma \tag{9.38}$$

where λ is the coefficient of dynamic friction of the fracture surface, i.e. the sum of the product of the effective stress and undamaged and the product of the fracture that has produced damage and the coefficient of dynamic friction and the actual stress is the actual stress.

Equation (9.38) can be rewritten as

$$\sigma_{eff}^n = \frac{1 - \lambda D}{1 - D} \sigma \tag{9.39}$$

After sorting, it gives the following equation:

$$\sigma_{eff}^n = \frac{A - \lambda d}{A - d}\sigma \qquad (9.40)$$

Bringing in the effective stress, the fracture expansion can be expressed as

$$d = \mu_d \int \sigma_{eff}^n \left\{ a - \frac{bck\sigma_{eff}^m \left(t + \sigma_0^{-*}\right)^{k-1}}{\left[1 + c\sigma_{eff}^m \left(t + \sigma_0^{-*}\right)^k\right]^2} \right\} dt + d_0 \qquad (9.41)$$

In summary, considering the formation of fissures, the deformation of rock salt in the third stage can be calculated by the following equation

$$\varepsilon = \int \sigma_{eff}^n \left\{ a - \frac{bck\sigma_{eff}^m \left(t + \sigma_0^{-*}\right)^{k-1}}{\left[1 + c\sigma_{eff}^m \left(t + \sigma_0^{-*}\right)^k\right]^2} \right\} dt + \varepsilon_0 \qquad (9.42)$$

(4) **Three-dimensional form**

The potential function is usually used to characterize the direction of plastic flow in plastic constitutive models [21]. If the potential function is similar to the yield function, it is called the correlation criterion. If it is different, it is called the noncorrelation criterion. In most cases, the flow direction measured in the experiment does not have a strong relationship with the yield function. Therefore, the non-associated criterion is used here.

$$\varepsilon_{ij} = \gamma \langle F \rangle \frac{\partial Q}{\partial \sigma_{ij}} \qquad (9.43)$$

where F is the yield function, the starting condition for creeping, Q is the plastic potential function, and its partial derivative with respect to time represents the plastic flow direction, γ is the relationship between creep deformation and stress, which can be calculated with Eq. (9.33). Theoretically, this model has a particular proportional relationship between the plastic deformations in each direction. However, most experimental results do not reflect this relationship. The accumulation of NaCl polycrystalline particles forms rock salt, and its deformation is mainly dominated by dislocation generation and slip. Therefore, the direction of the slip is related to that of plastic flow.

When the rock salt is in steady-state creep, the internal dislocation and external forces reach a dynamic equilibrium. The generation, slip and removal of dislocations occur within an orderly and stable state. The dislocation mainly slips along the direction of the system that most easily slips, which is the direction of the maximum shear stress, and the resulting plastic deformation is similar in all directions. In the first stage, the external force is obviously more significant than the internal force, and the dislocation multiplies and slips rapidly. In the rapid generation and slip process, the "supersaturation" state of dislocations may occur, resulting in the emergence of new slip systems, such as climbing and other behaviours, thus causing changes in the proportions of plastic deformation in all directions. Therefore, a "double potential function" constitutive model of the plastic potentials of the two stages was established in this paper.

$$\varepsilon_{ij} = \gamma_1 \langle F \rangle \frac{\partial Q_1}{\partial \sigma_{ij}} + \gamma_2 \langle F \rangle \frac{\partial Q_2}{\partial \sigma_{ij}} \tag{9.44}$$

The viscoelastic–plastic constitutive equation is

$$\varepsilon_{ij} = \gamma_1 \langle F \rangle \frac{\partial Q_1}{\partial \sigma_{ij}} + \frac{\Delta \sigma}{E} + \gamma_2 \langle F \rangle \frac{\partial Q_2}{\partial \sigma_{ij}} + \frac{\Delta \sigma}{E} \tag{9.45}$$

where γ_1 and γ_2 are the motion functions of the first and second stages, respectively, Q_1 and Q_2 are the plastic potential functions of the first and second stages, respectively. F is the start condition shared by the first two stages, and E is the elastic modulus. Here, a form similar to the Drucker–Prager (D-P) criterion was adopted by the plastic potential function.

$$Q = \sqrt{J_2} - \alpha I_1 \tag{9.46}$$

The derived creep–fatigue model for rock salt uses only eight parameters, which is not only able to consider the interaction between creep and fatigue, but also greatly reduces the parameters compared with other models. In the following, the derived rock salt creep–fatigue constitutive model will be used to validate the experimental data and analyze the applicability of the model.

9.2 Validation of the Creep–Fatigue Constitutive Model for Rock Salt in Creep, Fatigue and Creep–Fatigue Test

9.2.1 Validation of the Creep–Fatigue Constitutive Model in Pure Creep Test for Rock Salt

(1) The creep deformation and fitting results under single stress

The test specimens, the specimen environment, and the test apparatus are the same as the previous test, the constant stress level creep test program is as follows: 1 kN/s rate of uniform loading to the design stress level, the stress level of 4 MPa, 8 MPa, 12 MPa, 16 MPa and 20 MPa, respectively, loading and unloading path shown in Fig. 9.2.

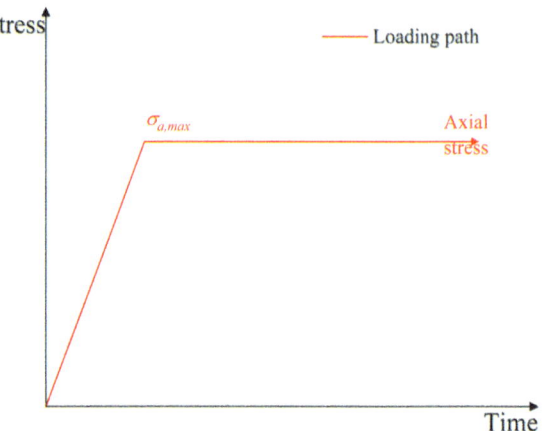

Fig. 9.2 Loading path for single stress level creep test on rock salt

The creep test curves were parametrically fitted by the creep–fatigue model, and compared with the Bailey-Norton model and the Burgers model, and the results are shown in Fig. 9.3.

In order to show the creep deformation process more clearly, Fig. 9.3 does not show the loading stage. The values of the vertical axis within the dashed circles represent the corresponding strain values when loading is completed. From the fitting results, compared with the Bailey-Norton model and Burgers model fitting results, the model can better fit the creep deformation under different stresses, including the decelerated creep stage and steady state creep stage. The five sets of fitted curves under different stress levels and different initial state variable levels all have a good fit with the experimental data and a small error, indicating that the model can better describe the creep characteristics of rock salt under different loads and has good applicability. The model parameters of the creep–fatigue model under different stresses are shown in Table 9.1. It can be found that the model parameters obtained change with the change of stress level, however, the regularity of the change is not obvious, which is similar to the disordered parameters obtained when fitting other models with different stress levels. This is mainly due to the fact that the multi-parameter model, in order to achieve the effect of the highest fit, ignores the overall optimal solution for some single parameters.

(2) The creep deformation and fitting results under increasing/decreasing stress

In order to verify the fitting effect of the creep–fatigue model on the creep deformation of rock salt under different loading and unloading paths and the same stress level, two sets of creep experiments of rock salt with elevated stress were carried out respectively. In the elevated stress experiments, the specimen was loaded to 4 MPa according to the loading rate of 1 kN/s and kept constant for 2 h. Subsequently, the elevated stress creep tests were conducted according to the same rate, kept for the same time, and increased by 4 MPa at each level until the specimen was damaged. The results of the increasing stress level creep test showed that the rock salt specimens

9.2 Validation of the Creep–Fatigue Constitutive Model for Rock Salt …

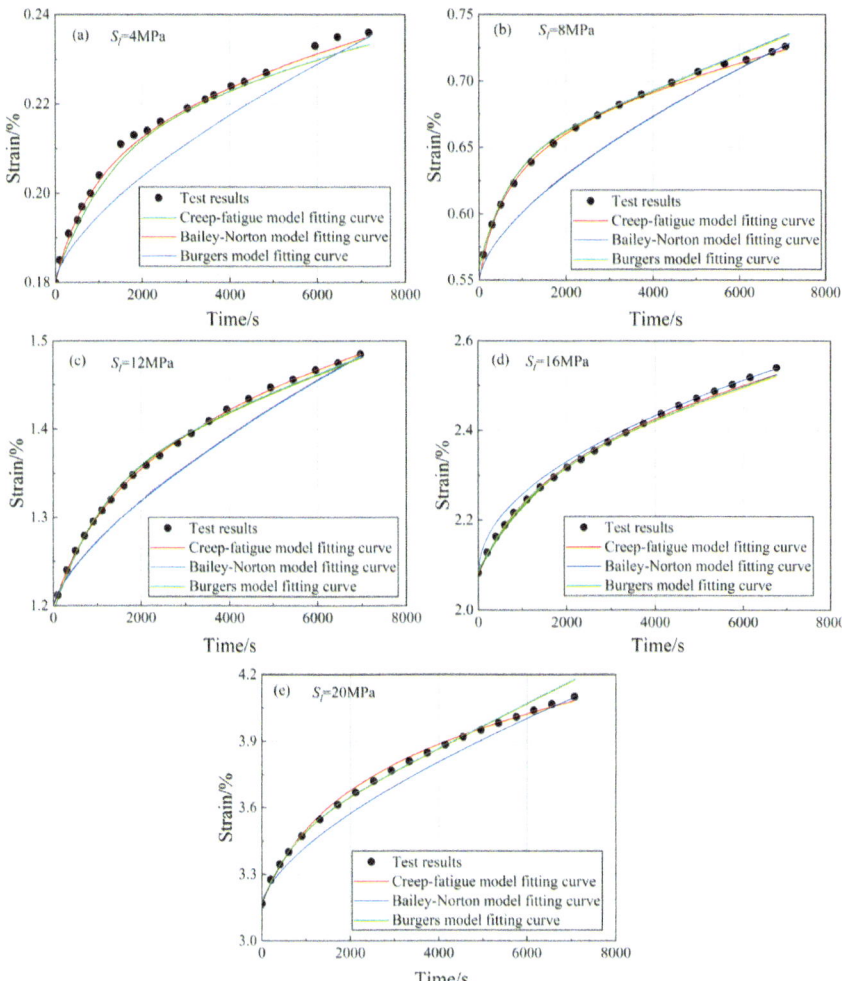

Fig. 9.3 Strain versus time for different model and experimental data, for creep tests with stress levels: **a** 4 MPa, **b** 8 MPa, **c** 12 MPa, **d** 16 MPa, and **e** 20 MPa on rock salt

Table 9.1 Parameters of creep–fatigue constitutive model

Stress/MPa	a	b	c	n	m	k	R^2
4	11	1.25	200	1.95	1	−0.72	0.99
8	11.5	1.05	42	2.13	0.7	−0.68	0.99
12	6.9	1.68	45	2.1	1	−0.61	0.99
16	6.9	1.4	45	2.1	1.2	−0.64	0.99
20	6.5	2	52	2.2	0.9	−0.7	0.99

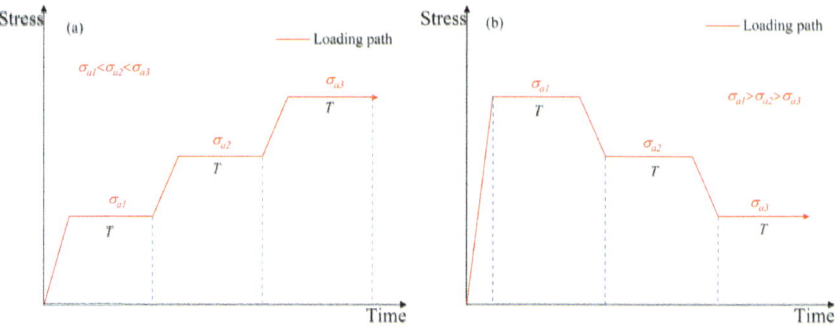

Fig. 9.4 Loading paths for **a** increasing stress level creep test and **b** decreasing stress level creep test

were damaged after the last step of 24 MPa, which was kept for 40 min only. Based on this, in the stress reduction experiment, the same loading rate was used to first load the rock salt specimen to 20 MPa, i.e., the pre-destruction stress level of the ascending stress creep, and kept for 2 h, with same rate used to reduce it to 0 step by step, with gradient of each step still 4 MPa. The loading and unloading paths are shown in the following Fig. 9.4.

The creep strain and creep rate at different stress levels of the increasing stress level creep test and the decreasing stress level creep test were compared and the results are shown in Table 9.2.

As can be seen from Table 9.2, the creep rate and the amount of creep of rock salt increase with stress levels in the increasing stress level creep tests. This phenomenon is similar to the results obtained in many literatures [22]. In the decreasing stress level creep test, both the creep rate and the creep amount of the rock salt gradually decrease with the decrease of the stress level. However, when the load was less than 16 MPa, the phenomenon of 'negative creep' appeared, which did not occur in the conventional ascending gradient loading. Comparing the creep tests of increasing stress level creep and decreasing stress level creep, the creep deformation rate of rock salts in decreasing stress level creep test is much smaller than that of increasing stress level creep test at the same stress level, except for the 20 MPa level. This is

Table 9.2 Comparison of test results

Stress/MPa	Increasing stress level creep test		Decreasing stress level creep test	
	Creep strain/%	Creep rate/h	Creep strain/%	Creep rate/h
4	0.057	0.0258	−0.012	−0.006
8	0.180	0.090	−0.002	−0.001
12	0.304	0.152	−0.003	−0.0015
16	0.460	0.230	0.029	0.0145
20	0.930	0.465	1.202	0.601

9.2 Validation of the Creep–Fatigue Constitutive Model for Rock Salt ...

mainly due to the fact that the internal structure of the rock salt in the increasing stress level creep test is gradually hardened under the action of the gradual increase of the external load (the degree of hardening and dislocation density are closely related, and will not be discussed again here); whereas, the 20 MPa in the decreasing stress level creep test is directly loaded to the corresponding level, and does not undergo the long-term hardening similar to that in the increasing stress level creep test, which results in the creep deformation rate under the maximum stress level in the decreasing stress level creep test is greater than that in the ascending gradient creep deformation rate. rate. In the decreasing stress level creep test, after the hardening effect of high stress, the creep deformation rate under the subsequent stress level is much smaller than that in the increasing stress level creep test, which is due to the fact that the hardening effect under the action of high stress is much larger than that produced by low stress.

The experimental results show that the stress loading path has a significant effect on the creep properties of rock salt. The experimental results in the increasing stress level creep test were validated using the creep–fatigue model, and the obtained fits and experimental results are shown in Fig. 9.5a.

The model fully considers the effects of time, load and state (loading stage and creep stage) on the creep characteristics, from the fitting results, it can be a better fit for both the creep stage and the loading stage, the 1st, 4th and 5th steps of the loading stage and the creep stage of the test curves and the fitted curves of the fit curve is higher, the 2nd and 3rd steps of the loading stage and the creep stage of the test curves and the fitted curves are basically the same. Although there is some error in the fitted curves, the change trends of the fitted curves and the test curves are basically the same. When the load was increased from 20 to 24 MPa, the rock salt showed an accelerated creep stage, and the specimen was destabilized and damaged, and the model showed good results for the simulation of the three stages.

For the decreasing stress level creep test of rock salt, the creep–fatigue model was validated against the experimental results of reduced gradient, and a comparison of the fitted situation and the test results obtained is shown in Fig. 9.5b. The rock salt

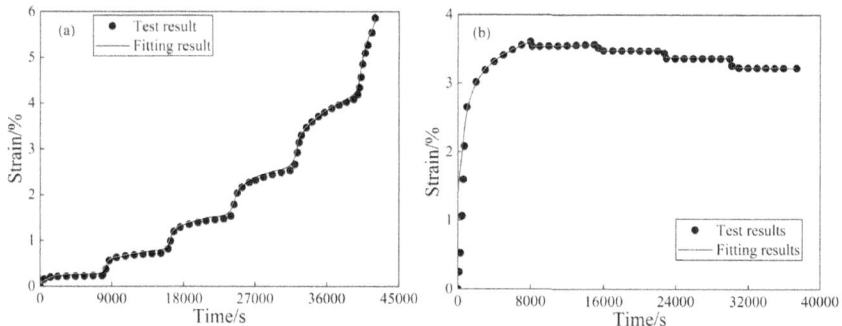

Fig. 9.5 Comparison of creep test results and creep–fatigue model fitting results for stress-leveling creep under **a** increasing stress and **b** decreasing stress

specimens were subjected to graded creep with a falling gradient, and the strains under all levels of stress levels produced both variable-rate creep strains and steady-state creep strains.

The model can not only fit the creep deformation law of the first stress level (20 MPa) well, but also can fit the creep deformation that occurs at each stress level after the reduction of the stress level, especially the phenomenon of 'negative creep' also has a good effect. 4 unloading stage and creep stage of the test curves and the fitted curves of the fit curve are high, and the error is very small. In particular, the first and fourth two stages of the test results and fitting results completely overlap, while the second and third stage of the test results and fitting results, show a small error of 2% only, indicating that through the introduction of state variables and variable-speed creep and steady-state creep unloading factor of the improved model and the degradation of decreasing stress level creep test data with a high degree of agreement.

Overall, the creep–fatigue model is able to well predict the development of creep deformation of rock salt under different lifting gradient stress paths. Compared with the traditional creep constitutive model, the model parameters change when the stress occurs, and the applicability of the model is greatly reduced, which brings a lot of constraints to engineering applications. The model only uses a set of parameters to completely describe the behavior of creep properties under different stress paths, and most of these parameters are only related to the material properties, which provides a convenient engineering stability assessment and application.

9.2.2 Validation of the Creep–Fatigue Constitutive Model in Pure Fatigue Test for Rock Salt

(1) The fatigue deformation and fitting results with pure fatigue test.

The data used in the test come from the rock salt pure fatigue test completed by our group in 2017 [23]. The loading and unloading scheme of the rock salt uniaxial fatigue test is as follows: the upper and lower limits of the test set stress level are 90% and 20% of the uniaxial compressive strength of the rock salt, respectively. The loading and unloading rate are set to 2 kN/s. The stress path is shown in Fig. 9.6a. According to the experimental results, the stress–strain curve is obtained as shown in Fig. 9.6b.

The fatigue deformation of rock salt and the approximation of rock rheology under static force are divided into three phases, which are the initial will phase, the isokinetic phase, and the accelerated phase. The loading phase represents the process of increasing stresses applied before the start of the cycle. In Fig. 9.6b, the hysteresis loops show a sparse-dense-sparse distribution with different phases, corresponding to the three stages of fatigue deformation, respectively, and the hysteresis loops are decreasing in interval in the initial phase, the interval in the isokinetic phase is basically the same, and the accelerated phase increases until the specimen is damaged.

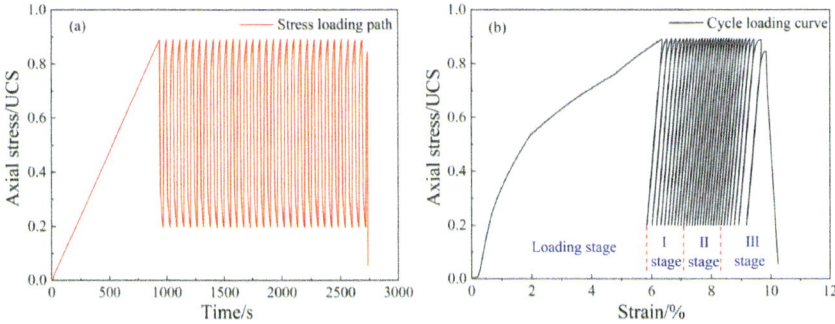

Fig. 9.6 Rock salt fatigue test: **a** stress path and **b** stress–strain curve

Using the creep–fatigue model proposed in this paper to fit the fatigue test data, the results are shown in Fig. 9.7, and it can be seen that the overall prediction curve of the model and the experimental results are very close to each other, which indicates that the prediction accuracy of the model is high. However, the fitting results in the loading stage differ greatly from the experimental data, and the reasons for this are analyzed as follows: the loading stage in the experimental process gradually increases from zero load, and under the influence of the compression and density of the primary fracture of the specimen, the specimen produces a large deformation, but due to the fact that in the operation of the air storage, the storage enclosing rock has often already gone through the initial loading stage, and thus the creep–fatigue model used is mainly to describe the deformation of the specimen after going through the loading stage. The model is mainly used to describe the deformation after the loading stage into the deceleration deformation or stable deformation stage, so the fitting effect for the initial loading stage is not good. For comparison, the prediction curves were shifted upward to coincide with the start of unloading in the first cycle (discussed later).

The test results of the time-strain relationship are also divided into three stages, as shown in Fig. 9.8. According to the comparison between the model prediction results and the experimental results, it can be found that the strain rate calculated by the model in the high stress part of the initial and accelerated phases is relatively low, and in the isotropic phase, the model calculation results are in good agreement with the experimental results. On the whole, the model can fully consider the effects of time, load and state on the fatigue deformation, and can well reflect the three stages of uniaxial compression cyclic loading and deformation of rock salt, with good applicability. In the three stages, the model prediction curves and the experimental curves almost completely coincide.

It should be pointed out that, in the last cycle of destruction, the specimen in the model can continue to carry the load, showing that the strain decreases with rebound in the process of stress decrease, which is mainly due to the fact that the actual testing machine is not a rigid testing machine, and the accumulated elastic energy of the testing machine is quickly released when the specimen is about to be

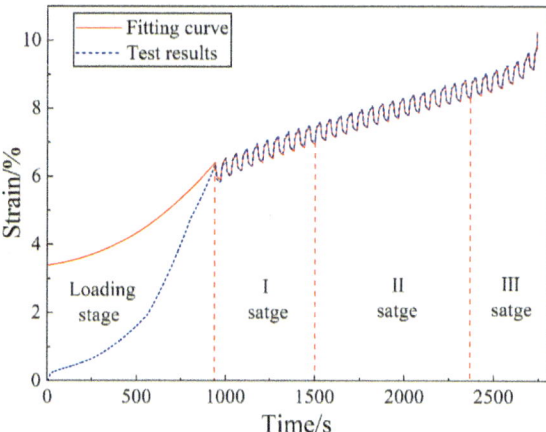

Fig. 9.7 Comparison between experimental results and fitted results, obtained from stress upper limit of 90% and lower limit of 20%

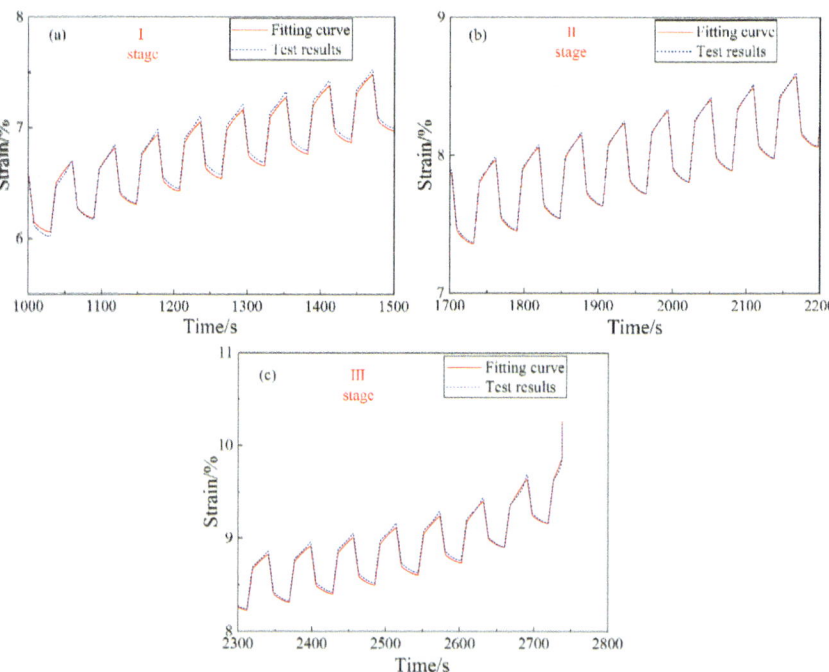

Fig. 9.8 Curve presented in Fig. 9.7 close-up on: **a** decelerating deformation stage, **b** steady deformation stage, and **c** accelerating deformation stage

destroyed, resulting in the rapid development of the cracks in the specimen, and thus the destruction. As a result, the experimental strain of the last cycle and the strain development trend predicted by the model differed greatly.

In order to verify the fitting effect of the creep–fatigue model on the fatigue deformation of rock salt under different stress ratios, cyclic loading and unloading fatigue experiments were carried out on the rock salt, with different lower and upper stress limits set. Divided into two groups of experiments, the first group, set the upper stress limit to 90%, the lower limit is set to 30%; the second group, set the lower stress limit to 20%, the upper stress limit is set to 95%. The obtained experimental results and fitting results are compared to those shown in Fig. 9.9a, b, and the model parameters are obtained as shown in Table 9.3.

According to the test data, when the stress ratio of uniaxial cyclic loading and unloading experiments is different, the fatigue life of rock salt and also corresponds to different, and the fatigue life increases with the increase of the lower or upper limit stress ratio. From the fitting results, it can be seen that the model is able to predict the different stages of uniaxial cyclic loading compressive fatigue deformation of

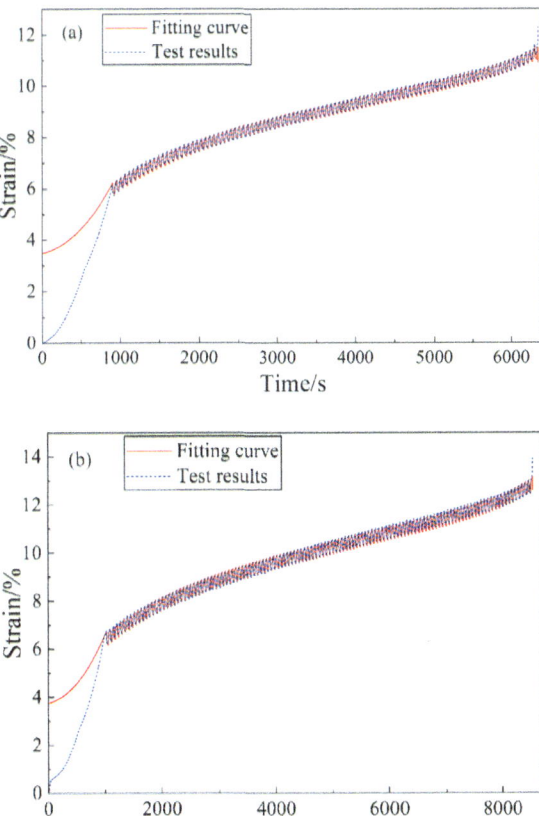

Fig. 9.9 Comparison graphs of test results and fitting results, **a** with stress upper limit at 90% and stress lower limit at 30%, and **b** with stress upper limit at 95% and stress lower limit at 20%

Table 9.3 Parameters of fatigue test model

Upper limit	Lower limit	a	b	c	d_0	μ_d
90%	30%	0.24	37	265	0.039	0.000015
95%	20%	0.30	37	340	0.034	0.00001

the results better, whether it is different upper limit stress ratios or lower limit stress ratios, and the prediction results are able to follow the stage characteristics better (i.e., higher degree of overlap). Similarly, the error in the first cycle loading stage comes from the compression-tightening effect of the first loading process; the deviation of the displacement trend in the final cycle is mainly due to the non-ideal rigidity of the rigid testing machine.

Overall, the model reflects the three stages of rock salt fatigue deformation well. The model considers the effects of time, load and state on rock salt fatigue, and the prediction/fitting results show that the simulation can fit the three stages under rock salt loading better. It shows that the model can describe the fatigue deformation of rock salt better and can reflect the three stages of fatigue deformation of rock salt well.

(2) The fatigue deformation and fitting results with low-stress interval fatigue test.

Again, the creep–fatigue model is validated upon the low-stress interval fatigue test of rock salt. The data is from Chap. 4.

The comparative graphs of the test results and model results are shown in Fig. 9.10, and the model parameters are obtained as shown in Table 9.4.

From the comparison between the model results and the test results in Fig. 9.10, the model can better predict the deformation process of unloaded rock salt at the lower limit interval. Selecting a cycle for comparison, it is found that the strain is increasing in two consecutive cycles of loading and unloading test results, and there is no obvious change in the strain of rock salt when it is in static stress, which is presumed to be due to the smaller constant force load (1 MPa) and shorter time, and the deformation generated is almost negligible.

Overall, the model has high consistency in the overall prediction of cyclic loading and unloading at different lower stress intervals. The model is able to reflect the mutual influence of creep fatigue deformation in the deformation process. From the model prediction results, it is seen that the model can fit the deformation under the creep–fatigue interaction of rock salt better, indicating that the model has good applicability.

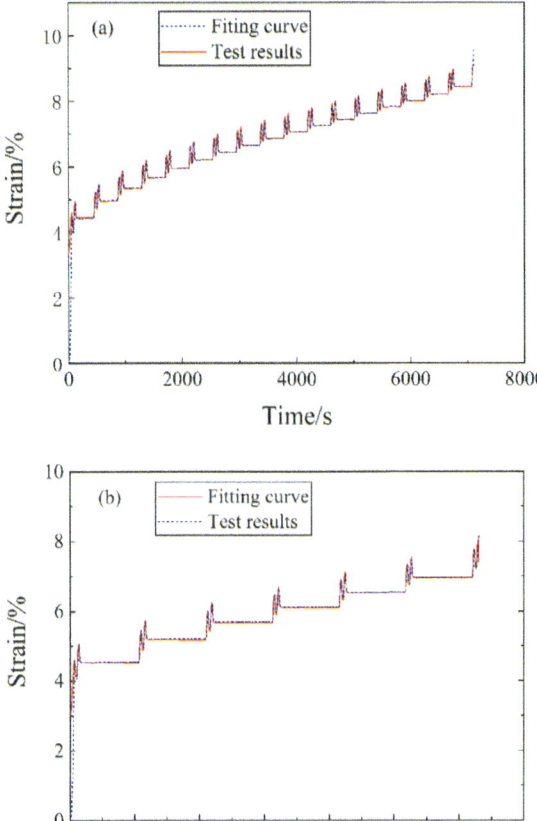

Fig. 9.10 Comparison of test results and fitted results, **a** low stress interval of 5 min, **b** low stress interval of 15 min

Table 9.4 Parameters of creep–fatigue model

Time/min	a	b	c	n	m	k
5	36	15.6	345	1.1	1.1	−0.72
15	25	15.0	280	0.9	0.92	−0.72

9.2.3 Validation of the Creep–Fatigue Constitutive Model in Creep–Fatigue Test for Rock Salt

(1) Firstly, the creep–fatigue model is verified on the creep–fatigue test completed in this thesis (Chaps. 5, 6 and 7), and the comparison are shown in Fig. 9.11.

Figure 9.11 shows the results of the three sets of creep–fatigue tests and the fitting results of the intrinsic model. Overall, the creep–fatigue model fits the results of the uniaxial, graded and triaxial creep–fatigue tests of rock salt with high accuracy,

Fig. 9.11 Comparison graphs of test results and fitting results, **a** uniaxial 5-min high-stress plateau creep–fatigue test (CCFT), **b** uniaxial 5-min graded high-stress plateau creep–fatigue (USCF) test, and **c** triaxial creep–fatigue test (TCFT) at a confining pressure of 3 MPa

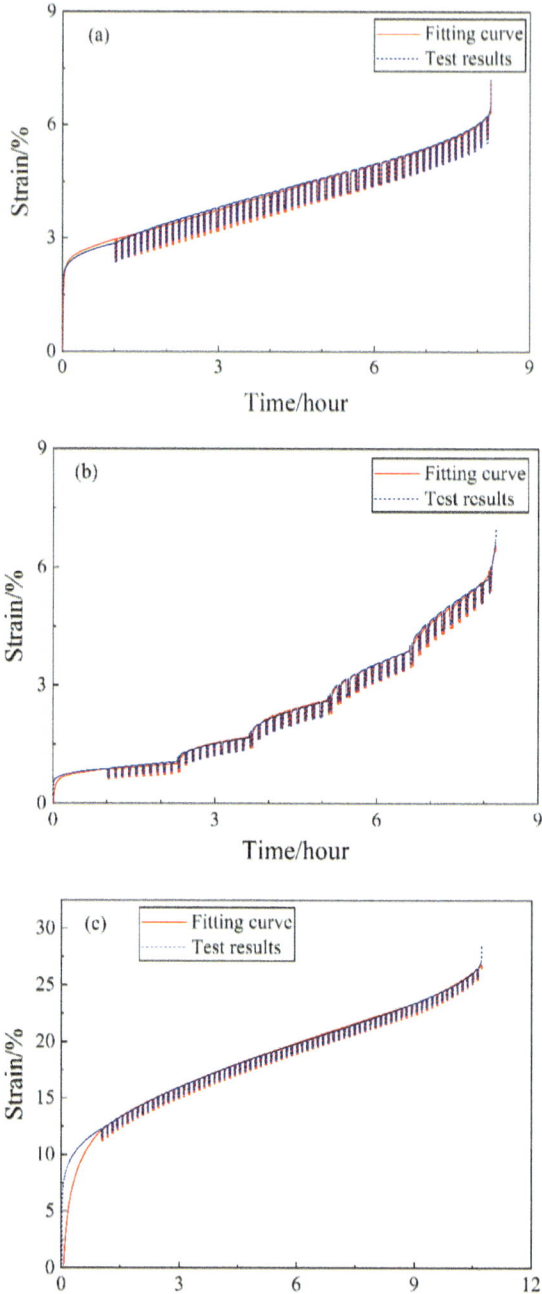

9.2 Validation of the Creep–Fatigue Constitutive Model for Rock Salt ...

which reflects that the proposed model has a good general applicability. Especially for the initial loading section, the fitting effect is also significantly improved compared with the pure fatigue test, which may be attributed to the following reasons: firstly, the modeling is based on the consideration of the interaction between fatigue cyclic load and constant stress creep load, so the fitting effect is naturally better for the creep–fatigue test, and secondly, the initial high stress plateau is added to the creep–fatigue test, which makes the test already in creep condition when it enters into cyclic load state, and this is the reason for the good fitting results of both graded and triaxial salt-rock tests. Secondly, the initial high stress plateau is added to the creep–fatigue test, so that the test is already in the creep state when entering the cyclic loading state, which is also closer to the actual state of the salt cavern surrounding rock, i.e., before the construction of the CAES plant, the salt carven surrounding rock has been subjected to a longer stage of bias stress load, and it has been in the process of rheology. For the graded creep–fatigue test, the proposed creep–fatigue model realizes that the characteristics of the creep–fatigue strain curves of rock salt under different stress paths can be perfectly fitted using one set of parameters, which provides a convenient way to assess the stability of the salt cavern CAES plant under different operating conditions during its operation.

(2) Secondly, the creep–fatigue model is verified on the long term creep–fatigue test completed in this thesis (Chap. 8), and the comparison are shown in Fig. 9.12.

Figure 9.12 shows the experimental and constitutive model fitting results for three sets of rock salt long-time creep–fatigue tests. Similarly, on the whole, the creep–fatigue constitutive model has better fitting results for the rock salt long-time creep–fatigue tests with different loading cycles and different stress ceilings, and the fitting accuracy is high, which reflects that the proposed model has good applicability for the long-time tests close to the operating frequency of CAES plants, and verifies the prospect of the model in engineering applications. From Fig. 9.12a cycle 8-h stress upper limit 18 MPa long-time creep–fatigue test and (b) cycle 24-h stress upper limit 18 MPa long-time creep–fatigue test, the two sets of tests. The creep–fatigue constitutive model can simulate the first two phases of the test better, even at very low loading and unloading stress rates. For the complete experience of three deformation stages of the cycle of 24 h stress upper limit 24 MPa long-time creep–fatigue test, creep–fatigue model for the accelerated deformation stage of the fit is also very good. It fully demonstrates that the model can be used to predict and evaluate the deformation and stability of the surrounding rocks of the salt cavern.

Fig. 9.12 Comparison graphs of test results and fitting results, **a** long-term creep–fatigue test with a stress upper limit of 18 MPa and a cycle duration of 8 h, **b** long-term creep–fatigue test with a stress upper limit of 18 MPa and a cycle duration of 24 h, and **c** long-term creep–fatigue test with a stress upper limit of 24 MPa and a cycle duration of 24 h

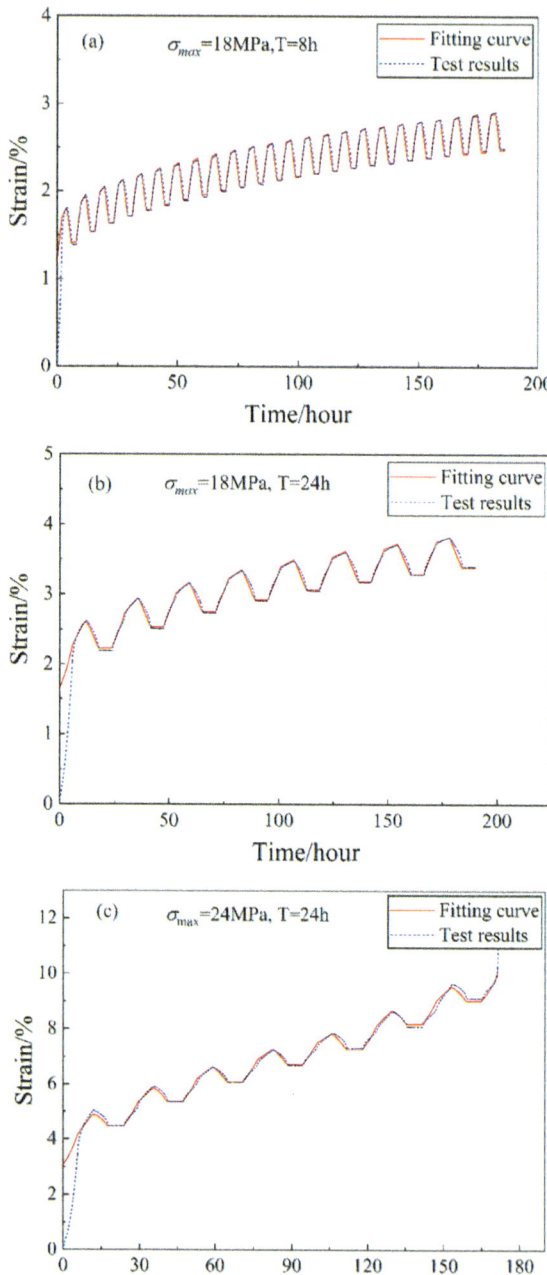

9.3 Analysis of the Influence of Model Parameters in Creep and Fatigue Tests for Rock Salt

9.3.1 Analysis of the Influence of Model Parameters in Creep Tests for Rock Salt

Comparison of model predictions and experimental results obtained from different stress ceilings as well as ascending and descending stress creep tests, it can be found that the established creep–fatigue damage ontological model is able to predict/describe the creep deformation behavior of rock salt under complex paths in a better way. The parameters involved in the model include a, b, c, n, m, k, d_0 and μ_d. In this part, the same stress path will be used, and the four parameters of c, n, m, and k, which have high influence in the creep test, will be selected to analyze and discuss their roles in the overall model.

(1) Analysis of the influence of k index on the model

Figure 9.13 shows the creep deformation-time diagram when the rest of the parameters are unchanged and k takes different values. As can be seen from the figure, when the k value is taken as − 0.2 and − 0.4, the rock salt creep process does not appear obvious non-stationary stage; when the k value continues to decrease, the creep curve will appear variable speed stage. And as the k value decreases, the faster the variable speed stage appears. k value is smaller, the same moment the creep strain is larger, at the same time, the variable speed creep stage of the creep curve radius of curvature is smaller, the earlier into the steady state creep. It can be seen that the change of k value has a greater impact on the rock salt creep model, and the size of k value directly affects the morphology of the creep curve of the rock salt, and the non-steady stage occurs when the k value < − 0.4.

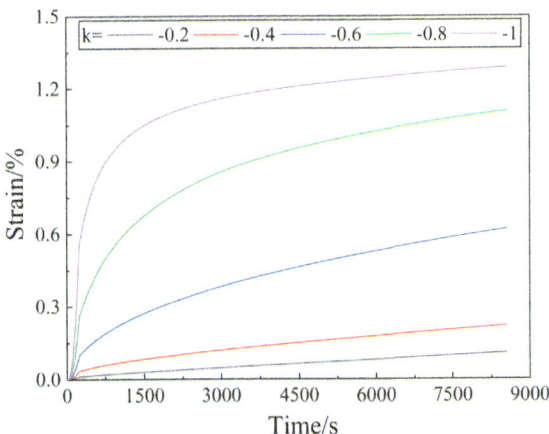

Fig. 9.13 Creep–fatigue curves under pure creep for different k values

The k metric affects the change in the creep curve by affecting the state variables in the model. Figure 9.14 shows a plot of the change in state variables when only the value of k in the model is changed. The creep curve plot and the state variable change plot have similar trends. As the value of k decreases, the curve gradually appears the stage of variable speed, the smaller the value of k, the faster the stage of variable speed appears, the larger the change of state variable, the earlier it tends to be stabilized; the larger the value of the state variable at the same moment, the higher the degree of hardening of the rock.

(2) Analysis of the effect of the m index on the model

When the stress and the rest of the parameters are the same, only change the value of m in the model. From Fig. 9.15, it can be seen that the larger the value of m, the larger the value of creep strain at the same moment, at the same time, the smaller the radius of curvature of the creep curve at the stage of variable-speed creep, and the earlier it enters the steady state creep. It can be seen that the size of the m value mainly affects the size of the creep strain and the length of the steady state creep stage.

Figure 9.16 shows the graph of state variables when changing the value of m in the model. The creep curve graph and the state variable change graph have a similar trend, the larger the value of m, the earlier the state variable value tends to stabilize; the larger the state variable value at the same moment, the higher the degree of rock hardening.

(3) Analysis of the effect of n index on the model

As can be seen in Fig. 9.17, when the value of n in the model is changed, the creep deformation decreases, and at the same time, the radius of curvature of the curve in the variable-speed creep stage becomes smaller, and the rate of change of the curve becomes slower and slower. It can be seen that the size of the n value mainly affects the size of the overall creep deformation amount and creep rate of the rock salt.

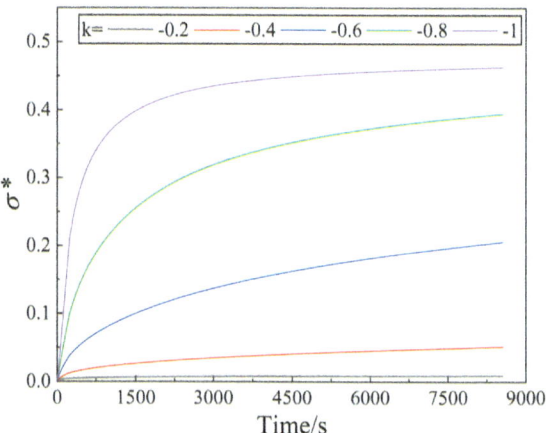

Fig. 9.14 The state-variable diagram of the model under different k values

9.3 Analysis of the Influence of Model Parameters in Creep and Fatigue … 203

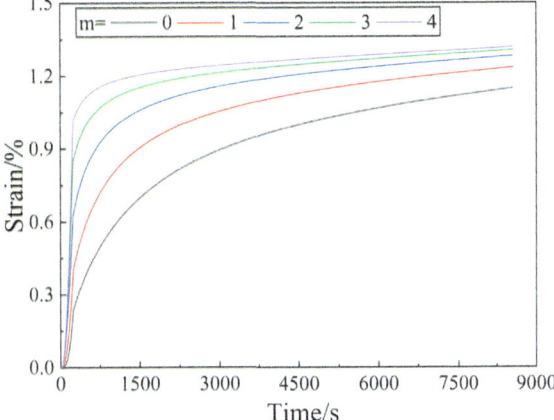

Fig. 9.15 Model creep curves under different m values

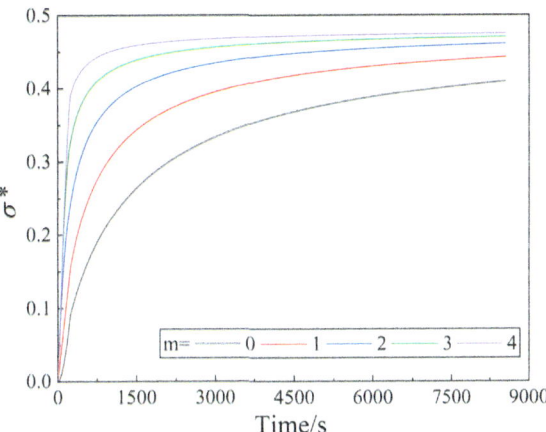

Fig. 9.16 The state variable diagram of the model under different m values

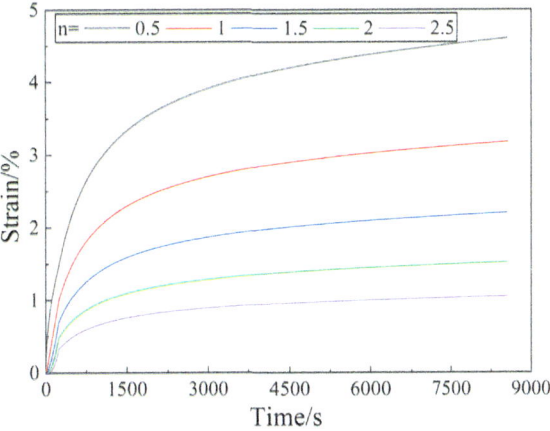

Fig. 9.17 Model creep curves under different n values

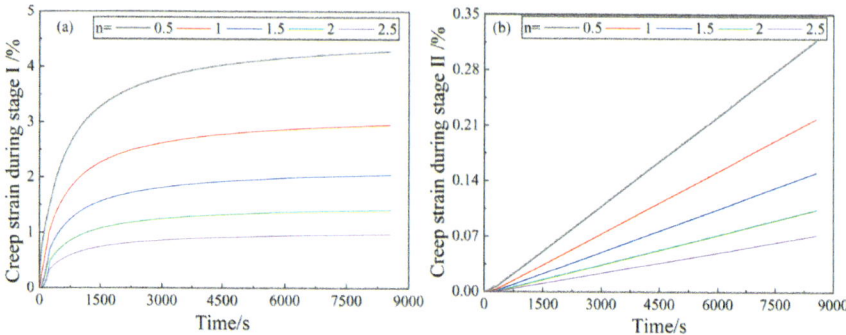

Fig. 9.18 **a** Model variable speed creep curves under different n values and **b** The constant velocity creep curve of the model under different n values

Since the accelerated creep phase is equally controlled by the two parameters d_0 and μ_d, and since the penetration and expansion of the cleavage surfaces are mainly generated during the loading and unloading process (especially in the third stage of the test). Therefore, for these two parameters will be discussed in the fatigue session. Therefore, the effects of model parameters on the decelerated creep and isokinetic creep stages are mainly analyzed here. n indicator affects both the change of decelerated creep strain in the model, characterizing the sensitivity of creep deformation to stress, as shown in Fig. 9.18a. It also affects the change in steady-state creep strain in the model, as shown in Fig. 9.18b. n values are larger, the smaller the values of both variable-velocity creep strain and isokinetic creep strain at the same moment. As the value of n increases, the variable-speed creep deformation is smaller and smaller, the radius of curvature of the curve is smaller and smaller, the creep rate change is slower and slower, and the creep rate is smaller and smaller; the larger the value of n, the smaller the isokinetic creep deformation is, and the smaller the creep rate is.

(4) c parameter of the influence of indicators on the model

Parameter a in the model is the overall coefficient of the model, which affects the overall creep deformation and rate; parameter b is the overall coefficient of the damping element 2, which affects the deformation and rate of deceleration creep. The significance of the more explicit, in the part of the creep test is not discussed for the time being. Keep the rest of the parameters unchanged and only change the value of c in the model. From Fig. 9.19, it can be seen that the smaller the value of c, the larger the value of creep strain at the same moment, and at the same time, the smaller the radius of curvature of the creep curve in the stage of variable-speed creep, the earlier the rock salt enters the steady state stage.

The c parameter affects the change of the creep curve by influencing the state variables in the model, as shown in Fig. 9.20. The creep curve graph and the state variable change graph have a similar trend. the smaller the c value, the faster the relative state variable changes in the variable speed creep stage, the sooner the state

Fig. 9.19 Model creep curves under different c values

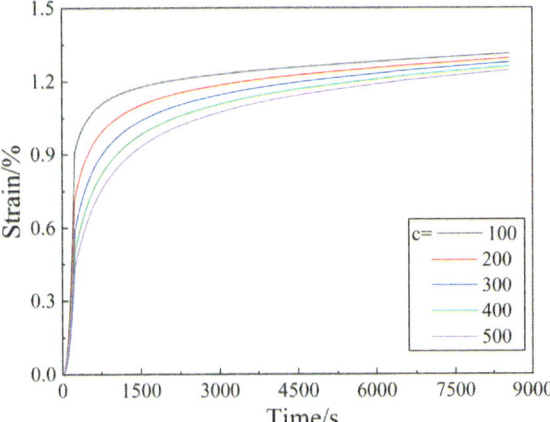

Fig. 9.20 The state variable diagram of the model under different c values

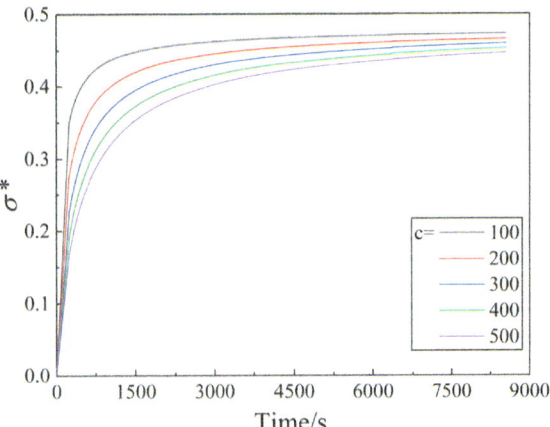

variable stabilizes; the larger the state variable value at the same moment, the faster of hardening.

9.3.2 Analysis of the Influence of Model Parameters in Fatigue Tests for Rock Salt

By changing different stress ratios to get the comparison of model prediction and experimental results, it can be found that the established creep–fatigue damage ontology model can better predict/describe the fatigue deformation behavior of rock salt under complex paths. In this part, the same stress path will be used, and the four parameters a, b, d_0, and μ_d, which have high influence on in the fatigue test, will be selected to act in the overall model to be analyzed and discussed.

Fig. 9.21 Loading path in fatigue testing of rock salts

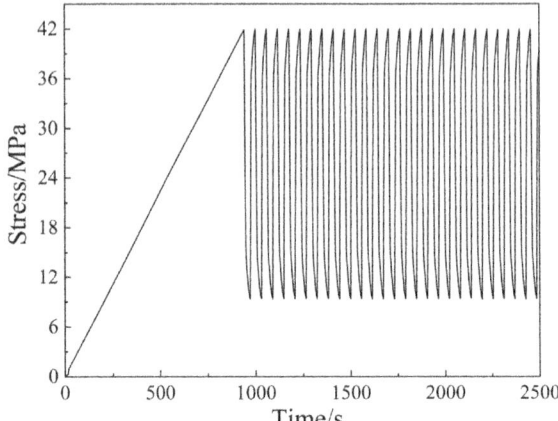

Rocks under uniaxial cyclic compression are divided into three stages, each of which is related to the loads applied to the rock. The deformation of rock salt is the sum of plastic and elastic deformation, and the elastic deformation is only related to the state of force, shown that the parameter change does not affect the elastic deformation change, and will not be described in the subsequent analysis. The stress path of the fatigue test is shown in Fig. 9.21.

(1) Influences on the steady state deformation phase (influence of a)

Only change the value of parameter a in the model, the rest of the parameters are kept unchanged, to get different fatigue curves as shown in Fig. 9.22. When $a = 0$, according to the formula can be seen, the steady state deformation rate of the material is also 0, there is no practical significance, but also the rock salt deformation is growing, so the value of a must be a positive number; as can be seen from Fig. 9.22, the first stage of deformation regardless of the size of a, basically overlap with each other, meaning that the value of a is almost independent of the decelerating strain of the first stage. The second stage of steady state (uniform) deformation stage, the mutual difference is larger, the overall slope of the curve in this stage with the increase in the value of a. The larger the value of a, the greater the rate of viscoelastic deformation of the rock; in terms of cyclic loading, the larger the value of a fatigue strain of a single cycle of the beginning and end of the difference between the value of a (i.e., residual strain) is also the greater, and therefore the greater the slope of the curve growth. As shown in Fig. 9.22b, fatigue deformation exhibits a similar pattern when elastic deformation is considered.

(2) Factors affecting the initial deceleration phase (effect of b)

When the other parameter values are kept constant, varying the size of the b-value, different fatigue strain component curves are obtained, as shown in Fig. 9.23. The relative differences between the fatigue curves with different b values are more significant, and the gap is more obvious from the initial stage to the steady state

9.3 Analysis of the Influence of Model Parameters in Creep and Fatigue …

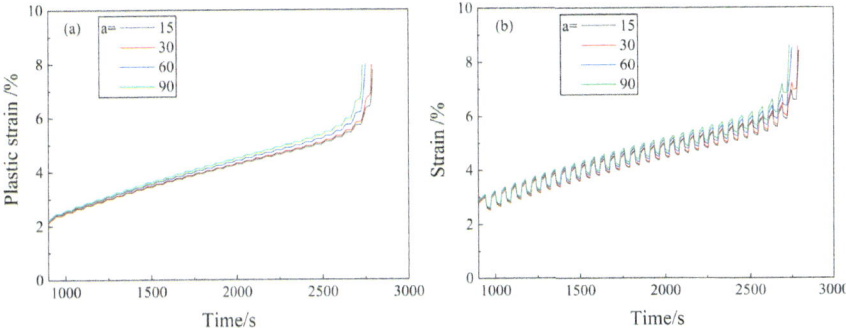

Fig. 9.22 Model deformation curves corresponding to different values of a **a** plastic strain and **b** total strain

stage. It can be found that when the b-value is small, the curve quickly enters the steady state deformation stage, and the overall development trend is almost linear; with the increase of the b-value, the curvature of the front development trend of the fatigue deformation curve increases, and the overall first stage (initial deceleration deformation stage) becomes longer, i.e., the duration increases. Therefore, the change of b-value only affects the variable speed strain part of the model prediction curve. As shown in Fig. 9.23a and b, as the value of b increases, and the corresponding radius of curvature at the beginning of the cyclic loading and unloading corresponds to a larger radius of curvature at the beginning of the cyclic loading and unloading, the corresponding deformation at the end of the loading phase increases with the value of b, thus affecting the overall rate of change of the model-predicted curves.

Overall, the parameter b mainly affects the variable-rate strain part of the fatigue deformation of the rock, which is mainly manifested in the radius of curvature at the beginning of the loading stage, and the size of the strain at the beginning of the initial stage, the larger the value of b is, the larger the radius of curvature is, and the

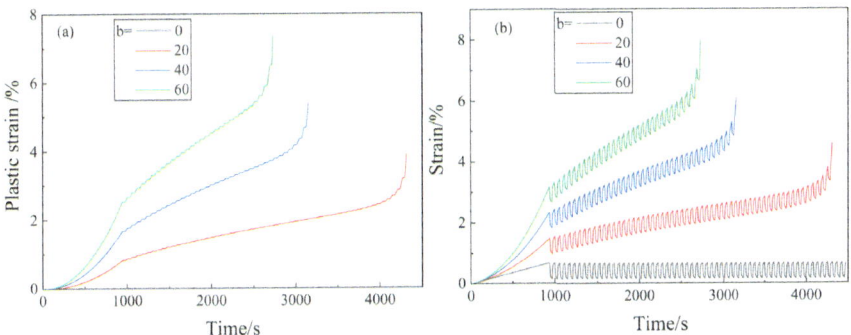

Fig. 9.23 Model deformation curves corresponding to different values of b **a** plastic strain and **b** total strain

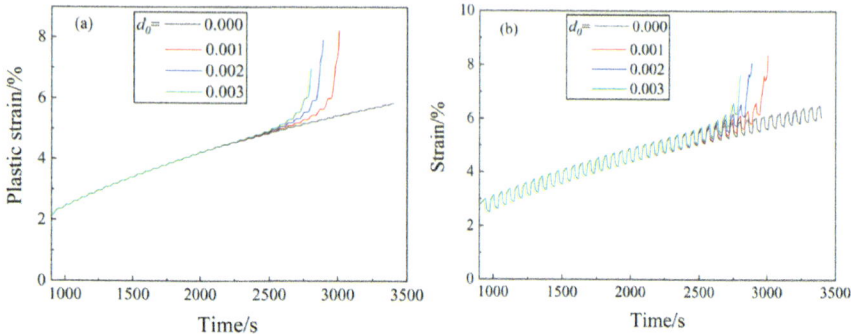

Fig. 9.24 Model deformation curves corresponding to different values of d_0 **a** plastic strain and **b** total strain

larger the initial strain is, which in turn affects the overall slope of the fatigue curve predicted by the model.

(3) Influences of deformation in the accelerated phase (effects of d_0 and μ_d)

When other parameters are unchanged, only the values of d_0 and μ_d in the model parameters are changed, and different fatigue strain component curves are obtained, as shown in Fig. 9.24 and Fig. 9.25. From the figure, it can be seen that the parameters d_0 and μ_d change, the initial stage of fatigue deformation of the rock and isotropic stage has almost no effect, different values of the corresponding curves almost completely overlap, only in the accelerated stage of the effect exists. d_0 increases, the curve in the accelerated stage of the increasing strain rate, combined with the model can be seen, near the destruction of the strain slope with the increase in the value of d_0 and gradually become larger. When μ_d increases, the strain rate of the curve in the accelerated stage is increasing, combined with the model, it can be seen that when the value of μ_d is 0–0.00008, the deformation in the stage of proximity to destruction increases slowly with the increase of the value of μ_d, and the overall change is very small, but when the value of μ_d is 0.00012, the strain is suddenly increased.

Overall, the parameters d_0 and μ_d together affect the variable-velocity strain and isochronous strain of the fatigue deformation, and mainly affect the accelerated strain phase in the model prediction curve.

9.4 Conclusions

In this chapter, based on a detailed analysis of the inadequacy of previous salt-rock creep models, the creep–fatigue intrinsic model of rock salt, which can consider the creep–fatigue interaction, was established by defining state variables and introducing unloading factor and fracture factor. Based on the mechanical test results of rock salt

9.4 Conclusions

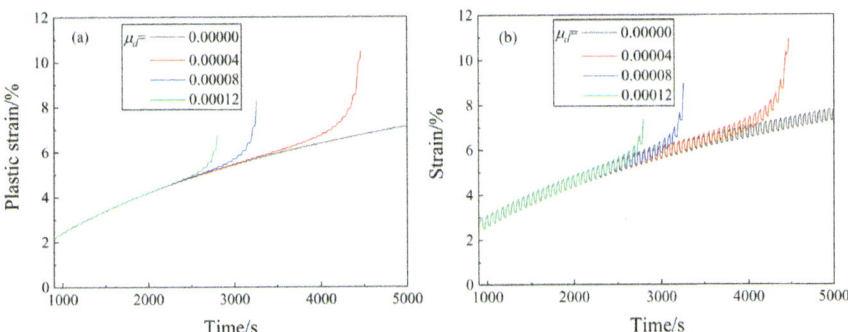

Fig. 9.25 Model deformation curves corresponding to different values of μ_d **a** plastic strain and **b** total strain

under three different stress paths of creep, fatigue and creep–fatigue, data analysis and model prediction are carried out to compare and analyze the accuracy of the model and thesss related parameter characteristics. The specific conclusions are as follows,

(1) A new creep–fatigue intrinsic model of rock salt considering creep–fatigue interaction is established by introducing the state variable characterizing the hardening degree of the rock on the basis of the Norton creep model. Based on the proposed creep–fatigue ontological model, the deformation of rock salt is divided into two parts: creep deformation (time-dependent deformation) and loading deformation (time-independent deformation).

(2) Four different loading paths, namely, constant stress creep test, cyclic loading and unloading fatigue test, uni/triaxial creep–fatigue test and long-time creep–fatigue test, are used to verify the creep–fatigue model. Comparison between the fitted curves of the four unused stress paths and the experimental curves shows that the agreement between the two is better, which means that the model can well describe the creep–fatigue model under different stress paths by taking into account the effects of time, loading and state on the creep and fatigue of the rock salt. characterize the creep–fatigue plastic deformation of rock salt under different stress paths.

(3) Through the analysis of the influence of different indexes on the model, the influence of different indexes on the creep and fatigue of rock salt is obtained. Parameter k mainly affects the decay rate of creep with time, which is the time-sensitive factor of creep rate decay; parameter m mainly affects the creep deformation amount and the length of the steady-state creep stage through stress, which is the stress-sensitive factor of creep rate decay; parameter n mainly affects the creep deformation amount and the size of creep rate through stress, which is the stress-sensitive factor of the model; Parameter c mainly affects the size of the amount of creep deformation and the length of the steady state creep phase, for the model of the overall strain rate decay sensitivity factor; Parameter *a* mainly affects the isotropic stage of the creep–fatigue damage model, and

the isotropic strain rate increases with the increase of the value of a. Parameter b mainly affects the initial stage of the creep–fatigue damage model, and the decelerating strain rate increases with the increase of the value of b, while the isotropic strain rate is almost unaffected by the value of b. Parameters d_0 and μ_d, on the other hand, jointly affect the acceleration stage of the fatigue damage model.

References

1. Zhen Yang, WanCheng Zhu, Kai Guan, et al. Influence of dynamic disturbance on rock creep from time, space and energy aspects[J]. Geomatics, Natural Hazards and Risk, 2022, 13(1): 1065–1086.
2. HaiYang Yi, Lele Lu, Wei Cao, et al. Parameters identification and comparative analysis of typical creep models of impurity salt rock[J]. Journal of North China Institute of Science and Technology, 2020, 17(2): 77–81.
3. TongBin Zhao, YuBao Zhang, QianQing Zhang, et al. Analysis on the creep response of bolted rock using bolted burgers model[J]. Geomechanics & engineering, 2018, 14(2): 141–149.
4. Shuquan Peng, Peiyu Wang, Ling Fan, et al. Research on elasto-plastic viscous fatigue constitutive model of jointed rock[J]. Rock and soil mechanics, 2021, 42(02): 379–389.
5. Liu Yang, Li Zhi-da. Nonlinear variation parameters creep model of rock and parametric inversion[J]. Geotechnical and Geological Engineering, 2018, 362985–2993.
6. Yonghui Li, Shaoyun Pu, Junying Rao, et al. Study on a modified Nishihara fatigue model for rock with cyclic loading[J]. Water Resources and Hydropower Engineering, 2017, 48(7): 129–135.
7. JingYang Ding, Hongwei Zhou, Qiong Chen, et al. Characters of rheological damage and constitutive model of salt rock[J]. Rock and soil mechanics, 2015, 36(3): 769–776.
8. Deyi Jiang, Wang Yifan, Wei Liu, et al. Construction simulation of large-spacing-two-well salt cavern with gas blanket and stability evaluation of cavern for gas storage[J]. Journal of Energy Storage, 2022, 48103932.
9. Yanlin Zhao, Ping Cao, Youdao Wen, et al. Elasto-viscoplastic rheological tests and nonlinear rheological modeling of rocks [J]. Chinese Journal of Rock Mechanics and Engineering, 2008, 27(3): 477–486.
10. Kaiyun Liu, Yongtao Xue, Hui Zhou. A nonlinear viscoelastic plastic creep model of soft rock with unsteady parameters[J]. Journal of China University of Mining and Technology, 2018, 47(04): 921–928.
11. YifanWang, Xiong Zhang, Deyi Jiang, et al. Study on stability and economic evaluation of two-well-vertical salt cavern energy storage[J]. Journal of Energy Storage, 2022, 56106164.
12. JunbaoWang, Xinrong Liu, Jianqiang Guo, et al. Creep properties of salt rock and its nonlinear constitutive model[J]. Journal of China Coal Society, 2014, 39(03): 445–451.
13. Fei Wu, Heping Xie, Jianfeng Liu, et al.Experimental study of fractional order viscoelastic-plastic creep modeling [J]. Chinese Journal of Rock Mechanics and Engineering, 2014, (5): 964–970.
14. Ruidong Peng, Zhide Wu, Hongwei Zhou, et al. A fine-scale experimental study of crack extension patterns in layered salt rocks [J]. Chinese Journal of Rock Mechanics and Engineering, 2011, (S2): 3953–3959.
15. Deyi Jiang, JinYang Fan, Jie Chen, et al. Compression-shear fatigue characterization and dislocation damage study of salt rocks [J]. Chinese Journal of Rock Mechanics and Engineering, 2015, (5): 895–906.

References

16. Masoud-K Darabi, Al-Rub Rashid-K-Abu, Little Dallas-N. A continuum damage mechanics framework for modeling micro-damage healing[J]. International Journal of Solids and Structures, 2012, 49(3-4): 492–513.
17. Lazar-M Kachanov. Rupture time under creep conditions[J]. International journal of fracture, 1999, 97(1-4): 11–18.
18. Fan Yang, Jinyang Fan, Zhenyu Yang, et al. Plasticity analysis and constitutive model of salt rock under different loading speeds[J]. Journal of Energy Storage, 2023, 67: 107583.
19. Qiang Liu, Yanlin Zhao, Liming Tang, et al. Mechanical characteristics of single cracked limestone in compression-shear fracture under hydro-mechanical coupling[J]. Theoretical and Applied Fracture Mechanics, 2022, 119103371.
20. A-G Evans. The role of inclusions in the fracture of ceramic materials[J]. Journal of Materials Science, 1974, 91145–1152.
21. Kavan Khaledi, Mahmoudi Elham, Datcheva Maria, et al. Stability and serviceability of underground energy storage caverns in rock salt subjected to mechanical cyclic loading[J]. International journal of rock mechanics and mining sciences, 2016, 86115–131.
22. Jie Chen, Chen He, Wu Fei, et al. Creep Properties of Mudstone Interlayer in Bedded Salt Rock Energy Storage Based on Multistage Creep Test: A Case Study of Huai'an Salt Mine, Jiangsu Province[J]. Geofluids, 2022, 2022.
23. Jinyang Fan. Fatigue damage and dilatancy properties for salt rock under discontinuous cyclic loading[D]. Chongqing university, 2017.

Open Access This chapter is licensed under the terms of the Creative Commons Attribution-NonCommercial-NoDerivatives 4.0 International License (http://creativecommons.org/licenses/by-nc-nd/4.0/), which permits any noncommercial use, sharing, distribution and reproduction in any medium or format, as long as you give appropriate credit to the original author(s) and the source, provide a link to the Creative Commons license and indicate if you modified the licensed material. You do not have permission under this license to share adapted material derived from this chapter or parts of it.

The images or other third party material in this chapter are included in the chapter's Creative Commons license, unless indicated otherwise in a credit line to the material. If material is not included in the chapter's Creative Commons license and your intended use is not permitted by statutory regulation or exceeds the permitted use, you will need to obtain permission directly from the copyright holder.

Chapter 10
Conclusion

10.1 Main Conclusions and Innovations

Due to the excellent rheological properties, low porosity, low permeability, and self-healing characteristics of rock salt, utilizing salt caverns for compressed air energy storage (CAES) is an effective approach to improving the efficiency of renewable energy utilization. Considering the actual operating conditions of CAES systems, the surrounding rock of the salt cavern is subjected to discontinuous cyclic loading under different gas injection frequencies and pressures, leading to alternating creep–fatigue effects. This book combines theoretical analysis, experimental research, and model derivation to investigate the following aspects of rock salt mechanics under various conditions: Mechanical and damage characteristics under monotonic loading, Creep mechanical behavior under different stress levels, Fatigue failure characteristics under different loading rates, Fatigue mechanical properties under different low-stress intervals, Creep–fatigue mechanical behavior under varying high-stress interval durations, Creep–fatigue mechanical properties under different confining pressures. Additionally, acoustic emission devices were utilized to monitor and analyze the effects of stress levels on the evolution of creep–fatigue damage in rock salt. Long-term creep–fatigue experiments were conducted on rock salt at actual frequencies consistent with CAES operation. The reasons and patterns underlying the interaction of creep and fatigue under various conditions were analyzed. Based on these interactions, a state-variable-based creep–fatigue constitutive model for rock salt was proposed and validated. The key innovations as blew:

(1) Dislocation theory was applied to explain the mechanical behavior of rock salt under conventional loading, cyclic loading, and discontinuous cyclic loading. This provides a theoretical foundation for advancing rock salt structural research at meso- and micro-scales.

(2) The effects of stress levels and loading rates on the creep and fatigue mechanical properties of rock salt were studied. It was demonstrated that stress paths significantly affect rock salt deformation. A rate-dependent model for rock salt was

developed, categorizing deformation into creep plastic deformation and loading plastic deformation.
(3) The influence of no-stress and low-stress intervals during discontinuous cyclic loading on plastic development in rock salt was investigated. The critical role of residual stress in discontinuous fatigue was identified.
(4) The interaction characteristics of creep and fatigue in rock salt were clarified, and the microscopic mechanisms of this interaction were revealed. This laid a theoretical foundation for the development of creep–fatigue constitutive equations for rock salt.
(5) A new creep–fatigue constitutive model for rock salt was established, incorporating state variables to characterize rock hardening. This model accurately describes the influence of historical loading–unloading paths on the viscoplastic mechanical properties of rock salt and precisely predicts creep–fatigue plastic deformation characteristics under various stress paths.

10.2 Implications for Future Study

This book, set against the backdrop of the construction and operation of salt cavern storage facilities in China, systematically investigates the creep–fatigue mechanical properties of the surrounding rock of salt caverns under the influence of multiple factors. The research combines theoretical analysis, experimental studies, and model derivation, yielding several valuable findings. However, further in-depth research is needed in the following areas:

(1) Integration of the Creep–Fatigue Constitutive Model into Finite Element Software

While this study derived a creep–fatigue constitutive model for rock salt that accounts for creep–fatigue interactions based on experimental results, practical engineering applications require numerical simulations using finite element software. These simulations should be calibrated against real-world monitoring data. Therefore, secondary development of the creep–fatigue constitutive model within finite element software is a key focus for future research.

(2) Applicability to Chinese Geological Conditions

The experimental samples used in this study were taken from Pakistan's massive salt domes, which differ from China's geological conditions. Most rock salt in China is stratified, with impurities and interlayer characteristics distinct from the rock salt used in the experiments. Thus, developing a creep–fatigue constitutive model specifically for stratified rock salt is another priority for future research.

(3) Permeability Changes in Salt Caverns

The gas tightness of salt cavern storage is also a critical factor for the safe operation of compressed air energy storage (CAES) systems. Due to limitations in experimental

10.2 Implications for Future Study

equipment, this study did not simultaneously test changes in the permeability of rock salt during the experiments. Investigating the changes in rock salt permeability under alternating creep–fatigue loading conditions will be an essential focus of future research.

Open Access This chapter is licensed under the terms of the Creative Commons Attribution-NonCommercial-NoDerivatives 4.0 International License (http://creativecommons.org/licenses/by-nc-nd/4.0/), which permits any noncommercial use, sharing, distribution and reproduction in any medium or format, as long as you give appropriate credit to the original author(s) and the source, provide a link to the Creative Commons license and indicate if you modified the licensed material. You do not have permission under this license to share adapted material derived from this chapter or parts of it.

The images or other third party material in this chapter are included in the chapter's Creative Commons license, unless indicated otherwise in a credit line to the material. If material is not included in the chapter's Creative Commons license and your intended use is not permitted by statutory regulation or exceeds the permitted use, you will need to obtain permission directly from the copyright holder.

The manufacturer's authorised representative in the EU is Springer Nature Customer Service Centre GmbH, Europaplatz 3, 69115 Heidelberg, Germany. If you have any concerns regarding our products, please contact ProductSafety@springernature.com

Printed and bound by CPI Group (UK) Ltd, Croydon, CR0 4YY

26/03/2026

02078941-0001